621.

FEATURE INTERACTIONS IN TELECOMMUNICATIONS SYSTEMS

Feature Interactions in Telecommunications Systems

Edited by

L.G. Bouma and H. Velthuijsen

Sponsored by the *IEEE Communications Society*
in cooperation with *ACM SIGCOMM* and *PTT Research*

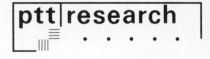

ComSoc

IOS Press
1994
Amsterdam • Oxford • Washington DC • Tokyo

© Koninklijke P.T.T. Nederland N.V., P.T.T. Research

All rights reserved. No part of this book may be reproduced, stored in a retrieval system, or transmitted, in any form or by any means, without the prior written permission from the publisher.

ISBN 90 5199 165 7
Library of Congress Catalog Card Number: 94-076399

Publisher:
IOS Press
Van Diemenstraat 94
1013 CN Amsterdam
Netherlands

Sole distributor in the UK and Ireland:
IOS Press/Lavis Marketing
73 Lime Walk
Headington
Oxford OX3 7AD
England

Distributor in the USA and Canada:
IOS Press, Inc.
P.O. Box 10558
Burke, VA 2209-0558
U.S.A.

Distributor in Japan:
Kaigai Publications, Ltd.
21 Kanda Tsukasa-Cho 2-Chome
Chiyoda-Ku
Tokyo 101
Japan

LEGAL NOTICE
The publisher is not responsible for the use which might be made of the following information.

PRINTED IN THE NETHERLANDS

Contents

Program Committee	VI
Introduction	VII

A feature interaction benchmark for IN and beyond
E.J. Cameron, N.D. Griffeth, Y.-J. Lin, M.E. Nilson, W.K. Schnure, and H. Velthuijsen 1

Using an architecture to help beat feature interaction
R. van der Linden 24

Towards automated detection of feature interactions
K.H. Braithwaite and J.M. Atlee 36

Classification, detection and resolution of service interactions in telecommunication services
T. Ohta and Y. Harada 60

Feature interactions among Pan-European services
K. Kimbler, E. Kuisch, and J. Muller 73

A building block approach to detecting and resolving feature interactions
F.J. Lin and Y.-J. Lin 86

Formalisation of a user view of network and services for feature interaction detection
P. Combes and S. Pickin 120

Specifying features and analysing their interactions in a LOTOS environment
M. Faci and L. Logrippo 136

Towards a formal model for incremental service specification and interaction management support
K.E. Cheng 152

Use case driven analysis of feature interactions
K. Kimbler and D. Søbirk 167

Interaction detection, a logical approach
A. Gammelgaard and J.E. Kristensen 178

Using temporal logic for modular specification of telephone services
J. Blom, B. Jonsson, and L. Kempe 197

The negotiating agents approach to runtime feature interaction resolution
N.D. Griffeth and H. Velthuijsen 217

Detecting feature interactions in the Intelligent Network
S. Tsang and E.H. Magill 236

Restructuring the problem of feature interaction: Has the approach been validated? Experience with an advanced telecommunications application for personal mobility
M. Cross and F. O'Brien 249

An architecture for defining features and exploring interactions
D.D. Dankel, M. Schmalz, W. Walker, K. Nielsen, L. Muzzi, and D. Rhodes 258

Author Index 272

Program Committee

Chair: E. Jane Cameron (Bellcore, USA)

Jan Bergstra (CWI and University of Amsterdam, The Netherlands)
Ralph Blumenthal (Bellcore, USA)
Kong Eng Cheng (Royal Melbourne Institute of Technology, Australia)
Bernie Cohen (City University of London, UK)
Fulvio Faraci (CSELT, Italy)
Robert France (Florida Atlantic University, USA)
Steve German (GTE Labs, USA)
David Gill (MITRE, USA)
Toru Ishida (Kyoto University, Japan)
Richard Kemmerer (UCSB, USA)
Eric Kuisch (PTT, The Netherlands)
Victor Lesser (University of Massachusetts, USA)
Yow-Jian Lin (Bellcore, USA)
Luigi Logrippo (University of Ottawa, Canada)
Robert Milner (BNR, UK)
Leo Motus (Tallinn Technical University, Estonia)
Jacques Muller (CNET, France)
Jan-Olof Nordenstam (ELLEMTEL, Sweden)
Stott Parker (UCLA, USA)
Ben Potter (BNR, UK)
Henrikas Pranevitchius (Kaunas University of Technology, Lithuania)
Lynne Presley (Bellcore, USA)
Jean-Bernard Stefani (CNET, France)
Greg Utas (BNR, Canada)
Juri Vain (Institute of Cybernetics, Estonia)
Yasushi Wakahara (KDD R&D Laboratories, Japan)
Ron Wojcik (BellSouth, USA)
Pamela Zave (AT&T Bell Laboratories, USA)

Introduction

The papers in this book, with one exception, were presented at the Second International Workshop on Feature Interactions in Telecommunications Systems (FIW '94) held in Amsterdam, 8–10 May 1994. The exception is "A feature interaction benchmark for IN and beyond", which appeared earlier in abbreviated form in the IEEE Communications Magazine in March 1993. The extended version, however, has never been made widely available. It is included in this book to serve as a tutorial to understand some of the intricacies of the feature-interaction problem and to help understand the relevance of the other contributions in this book.

FIW '94 is the second in a series: the first workshop was held in St. Petersburg, Florida, USA in December 1992. The mission of the workshops is to encourage researchers from a variety of fields in computer science (software engineering, protocol engineering, distributed artificial intelligence, formal techniques, software testing, and distributed systems, among others) to apply their techniques to the feature-interaction problem and to bring them together with designers and builders of telecommunications software.

Features are modifications of or enhancements to the control of telecommunications services. A feature interaction occurs when the behavior of one feature is changed by the behavior of another. In many cases this can lead to unexpected or undesired behavior which affects the quality of the service provided to telecommunications users. The feature-interaction problem is a major obstacle to the rapid deployment of new telecommunications services.

The feature-interaction problem is not unique to telecommunications software. It can be seen as a special case of problems that are encountered in the development and maintenance of large, distributed software systems that require frequent changes and additions to their functionality. Therefore, we trust that this book will be of interest to a broader audience than to just researchers in the telecommunications domain.

The need for an integrated approach

The ultimate goal of research in the area of feature interactions is to make sure that interactions are not a problem anymore: that users do not experience unexpected and undesired behavior of features, and that service and network providers can guarantee this at a reasonable cost. However, the feature-interaction problem is diverse and complex, as has been argued before in the literature (for instance, see [4]). The diversity of the problem suggests that there will not be a single technique or technology for solving it, at least not in the near future. Thus, we should look at what could be a useful combination of techniques that together would solve the feature-interaction problem in general or for a chosen service platform.

We can categorize different techniques according to how they address the feature-interaction problem. We distinguish techniques for avoiding, detecting, and resolving interactions. Within each category there may exist different approaches that supplement each other. An integrated approach to solving the feature-interaction problem will consist of techniques from each of the three categories, and within each category, of as many as are needed to obtain a full coverage of the problem.

We describe these categories in some more detail. The categorization is used below to introduce the contributions by papers in this book.

Avoidance. Approaches for avoidance of interactions are aimed at developing service platforms and service creation environments that lead to service implementations that are intrinsically less prone to interactions. Examples of such approaches are incorporation of open distributed computing platforms in telecommunications systems to deal with interactions that are termed *"intrinsic problems in distributed systems"* in [2], or definition of an extended signalling protocol between network and customer equipment to prevent interactions that occur due to limitations of existing signalling protocols (see also [2]). Additionally, guidelines for service creation and service management fall within this category as far as they address feature interactions.

Detection. Given that it is not very likely that all feature interactions can be prevented, especially not with current telecommunications systems and service-creation practice, they need to be detected. The cause of an interaction may be introduced at all points of the service-creation process as more detailed descriptions of a feature are generated (cf. the software life-cycle view to classifying interactions of Bowen et al. [1]). Thus, it may be opportune to check for interactions at various stages of the service creation process. It seems reasonable to require that all interactions are detected during the service creation process, i.e., off line and not when features are already operational. However, the number of services and service features involved, the number of ways in which features can interact, as well as the size of a complete, detailed feature implementation description will usually make the complexity of off-line detection of all possible interactions too high to be feasible. In those cases, on-line detection may be the answer because then one needs to consider only the instantiated parameters and invoked features in any specific situation. Examples of off-line detection mechanisms are verification and validation tools, simulation environments, and the use of checklists. An example of on-line detection is behavior analysis (see below).

Resolution. Once an interaction is detected, one must determine if the interaction is indeed harmful and how the interaction can be resolved. Similar to detection, resolution can be done both off-line and on-line. Off-line interaction resolution includes not only such activities as manual redesign or re-implementation, but also customer care. One should keep in mind that the feature-interaction problem occurs when features exhibit unexpected or undesired behavior. Thus, changing the expectations of users to match the actual behavior of features may as well resolve certain feature interactions as reimplementing the feature might do. On-line resolution approaches are those that find a way to resolve an interaction when it threatens to occur. Several on-line resolution approaches that have been proposed before are the feature-interaction manager (FIM) in IN, event-based resolution mechanisms, and negotiation (see below).

These three categories are shown in Figure 1. Furthermore, we distinguish between off-line and on-line techniques, except for avoidance, where this distinction makes less sense. For each of the resulting categories a few possible approaches are listed.

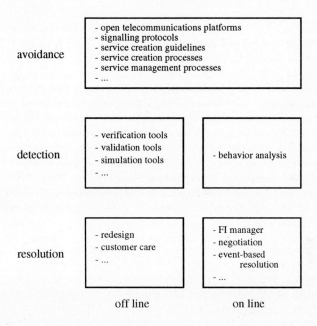

Figure 1: Categorization of approaches to addressing feature interactions.

Interestingly, there exists an abundance of approaches in the literature to off-line feature-interaction detection, while there exists hardly a well-documented, systematic approach to off-line feature-interaction resolution. Reversely, there exists hardly any approach dedicated to on-line detection, but several approaches to on-line resolution. This can be explained as follows. Off-line resolution is in general considered to be a knowledge-intensive and creative effort which is hard to capture by a systematic approach. On-line detection of interactions is simpler than off-line detection, because on-line detection typically has to deal with only the data relevant for a particular situation and not for all possible situations in which a certain pair of features could be involved. Thus, on-line detection is usually studied only as a subtopic of on-line resolution.

Contributions in this book

We mention the contributions of the papers gathered in this book briefly. Most of the papers dealing with off-line detection use a common approach. To avoid repetition, we summarize this approach here. Two descriptions are made for each feature: a high-level, behavioral one and a lower-level, procedural one. Interactions are detected by checking whether the combined procedural descriptions do not invalidate the combined behavioral ones. The languages used for the descriptions and the checking mechanism vary.

1. **A feature interaction benchmark for IN and beyond.** As stated before, this paper is in fact not a contribution to this conference, but appeared in shortened form in IEEE Communications Magazine in March 1993. It contains categorizations of

feature interactions and many examples, several of which re-appear in these proceedings.

2. Using an architecture to help beat feature interaction. This contribution by the invited speaker to the conference, describes how results of work on open distributed processing architectures appear to be applicable to the solution of certain feature-interaction problems in telecommunications systems. The principles of *separation* and *substitutability* are instrumental in application-independent structuring of systems. The principle of *separation* allows systems to be structured in ways which reduce the chance of feature interactions which are introduced due to unexpected resource sharing. The principle of *substitutability* allows explicit checks to be made whenever two software components are to mutually cooperate through a set of required interactions, thereby at least reducing and perhaps eliminating unexpected mismatches.

3. Towards automated detection of feature interactions. This contribution proposes to tackle the detection of interactions by describing telephone calls as layered finite-state machines. One starts with finite-state machines for the originating and terminating end of a call, and adds features as layers on top of these.

The authors introduce an informal graphical notation for these machines and a formal tabular notation, where states, input for and output of state transitions, and assertions are the primitive elements. Information between the different layers is passed as *tokens*. Algorithms are sketched where interaction is detected by tracing execution paths. This enables spotting several types of possible interactions: information interaction, call-control interaction, assertion interaction, and resource interaction. Four examples are given, based on informal reasoning about execution traces.

4. Classification, detection and resolution of service interactions in telecommunication services. Again, finite-state machines are used to describe behaviors of telephone systems. The approach sketched in this paper differs from the previous one in taking a more user-oriented view: the network is considered to be a black box, and service specifications are descriptions of desired terminal behaviors. Interactions are now modelled as unwanted properties of these *descriptions*. For example, a certain combination of finite-state machines can have deadlock states, or exhibit non-determinism, or lose transitions. The authors developed the language STR for describing finite-state machines. An interpreter to assist in the detection of inconsistencies in combined descriptions is under development.

5. Feature interactions among Pan-European services. The authors describe the analysis method developed in a Eurescom[1] project (PEIN) for detection of interaction between telephony services which are to be introduced in the European Community between 1996 and 1998. A significant aspect of this method is the division of features into categories, where two features are in the same category if they have similar functionality. Interaction appears then as an interaction between these categories. This drastically downsizes the complexity of the interaction-detection problem. A structured approach is sketched for detection of service interaction by breaking up services into their constituent features and concentrating on possible interaction between features from interaction-prone categories in different stages of their life-cycle.

6. A building block approach to detecting and resolving feature interactions. A description of a formal approach to the problem of interaction between IN-type features, again based on finite-state machines, is given. Features and their operating contexts are modelled as building blocks that can be combined into complete descriptions. The authors treat three operating contexts: originating side of the call,

---

[1] Eurescom is a joint research initiative of several European public network operators.

terminating side, and two-party call, and claim that these suffice to treat all relevant cases of interaction. Features have a procedural description (their finite-state machine) and assertions that form a behavioral description. Interaction occurs when the procedural description of combined features invalidates the behavioral description of at least one of them.

This abstract approach is detailed by describing an ongoing experiment at Bellcore where the finite-state machine descriptions are made using Promela, and assertions are done in a linear temporal language. For checking, these assertions are compiled into formulae that SPIN, the accompanying verification tool for Promela, can handle. Included in the experiment is the development of a tool that offers assistance when constructing formal descriptions for features.

7. Formalisation of a user view of network and services for feature interaction detection. This paper describes a line of research that is quite close to those of the previous one. It reports on work ongoing in the SCORE project of the RACE-II technology program. The description language chosen here is SDL, and the language used for the behavioral specification of features is a temporal logic, which might be linear or branching-time. An implementation route is sketched with an SDL tool set that soon will offer the possibility to compile temporal formulae into *observer* processes. These observers will then enable the automatic verification of correctness of composition of features against their behavioral specifications. The approach is illustrated with several examples of formal feature descriptions.

8. Specifying features and analyzing their interactions in a LOTOS environment. The use of a formal description technique is again advocated to define the interaction problem, and to come up with a detection method at the specification level. The approach addresses POTS features, and uses the language LOTOS. What makes this different from the two previous approaches, is that LOTOS is used for both behavioral and procedural descriptions. Several types of constraints are introduced following the knowledge-oriented model of Halpern and Moses[5]: local, end-to-end, and global constraints. Local constraints are modelled by using a constraint-oriented specification style in the processes that describe the separate features. Global constraints are enforced by introducing a new process that models these constraints, and that runs synchronized with the feature processes. An animation or verification tool is used to detect interaction within the composite specification. The approach is illustrated with two classic examples of interaction.

9. Towards a formal model for incremental service specification and interaction management support. This approach also uses LOTOS, but concentrates on the development of an IN-type model for the specification of features. Starting from a Basic Call Process derived directly from the ITU (formerly CCITT) Q12xy Recommendations for the IN CS-1 standard, specification of features as SIBs and a way to add them to this BCP using operators from LOTOS is described. A restricted type of interaction is solved by the introduction of an interaction manager that eliminates nondeterminism in the specification.

10. Use case driven analysis of feature interactions. The authors adapt a method – Use Case Driven Analysis – from the Object Oriented Software Engineering domain. This method starts from an informal, user-oriented description of a service by generating *use* cases, i.e., scenarios of system usage described as sequences of events and user interactions with the system. These use cases are then stepwise transformed into so-called *service usage models* describing the dynamic behavior of the system, still from the user point of view. The service usage models are analyzed for error-prone combinations of features, which are taken to be features that use the same data. The analysis phase is amenable to tool support and the paper finishes with the description of such a prototype tool. The method is related to the one from the PEIN project (see above).

11. Interaction, a logical approach. This paper describes another formal approach that is related to work of Inoue et al. [6] and Cameron and Lin [3], but introduces it differently. The basic models used by the authors are deterministic, labeled transition systems, over which *network* properties (properties valid in all states) can hold, and *transition rules* (in a somewhat unusual sense) can be fulfilled. Transition rule in this context means: if a state fulfills a (first-order) precondition, a certain transition is enabled, and the target state fulfills a corresponding postcondition. Interaction appears in this model as non-realizability of certain models. A Basic Call service is the core of all systems, on top of which feature specifications can be added. The method is illustrated with several examples.

12. Using temporal logic for modular specification of telephone services. In this approach, a central role is attributed to the language used for the specification of the behavior of features. The underlying model is a finite-state machine model, and the logic is inspired by Lamport's Temporal Logic of Actions. Linear temporal formulas are built from first-order formulas (with event predicates and action formulas describing state-space changes) using the linear temporal operator *always*. Interaction between features is modeled as above: the (behavioral descriptions of) two features interact when they result in an inconsistent system description. As the authors point out, the approach reminds one of adding a linear temporal logic to the Z specification language. The method is applied to several examples.

13. The negotiating agent approach to runtime feature interaction resolution. This is one of the few papers that addresses the issue of run-time detection and resolution of interactions. It builds upon previous results obtained by implementing a negotiating agent model on top of the Touring Machine running at Bellcore. The aim of the paper is to show that these results can be extended to implement negotiation on an arbitrary telecommunications platform if this platform has a set of well-defined operations on its interface. The approach is illustrated by a proposal for the implementation of example AIN and IN CS-1 features on the Touring Machine platform.

14. Detecting feature interactions in the Intelligent Network. This contribution also deals with detection of interactions in a run-time situation, but is more modest in its approach by only looking at detection. The context is an IN-type network, where features are instantiated together with a *Feature Manager* (FM) that checks if an activated feature displays correct behavior. This is done by requiring a feature to behave as a deterministic finite-state machine and to have the FM perform a behavior analysis and an analysis of the resources modified by the feature in each step. The approach has been validated on a simulated IN testbed with four IN features and multiple users.

15. Restructuring the problem of feature interaction: Has the approach been validated?. A drastically different point of view is introduced: communication through a telecommunications network can be seen as a form of social interaction. If one concentrates on the goal of the communication, then other activities can be introduced to realize this goal. In this sense, (unwanted) feature interactions can be avoided by restructuring the problem. An experimental tool has been developed to model communication that is able to find solutions to an interaction problem by manipulating parameters of time, location, and media. The authors report that, interestingly, the

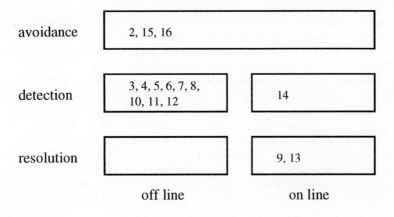

Figure 2: Contributions.

interaction problem reappears at a lower level (the level of representation of data structures).

16. An architecture for defining features and exploring interactions. The premise of this paper is that feature interactions occur often because their natural language descriptions are inconsistent and incomplete. The proposed remedy is a system that converts English-based requirement documents into a knowledge-based representation. The paper describes such a system (under development) that assists a specifier in the production of requirements. Its components are a natural language front-end, a command interpreter that updates a Knowledge-Data Base with predicates formed by the user interface, and the Knowledge-Data Base itself based on a newly developed language for knowledge representation. The paper concludes with a discussion on how this knowledge base can aid a specifier in discovering interactions.

Conclusions

We have mapped the contributions in this book onto the categorization of the previous section. The results are shown in Figure 2.

Considering the number of different approaches that can be found in, e.g., this book, in the August 1993 issues of Computer and IEEE Communications Magazine, and in the proceedings of various conferences and workshops, one sees a divergence of approaches rather than a convergence. We are excited to see that more and more new approaches are emerging. However, one also notices that there are not yet published many efforts exploring the present and potential scope of existing techniques or comparing techniques from the same category. Nor does one encounter descriptions and evaluations of combined approaches for dealing with feature interactions in day-to-day service provisioning. We hope to see progress particularly in these areas in the near future.

Acknowledgments. The editors would like to express their thanks to all people that have contributed to the appearance of this book. Specifically we would like to mention the often hard labor of the program chair Jane Cameron and all members of the program committee. Also, PTT Research management is gratefully acknowledged for their support, notably Jeroen Bruijning and Hans de Stigter. Finally, all this would not have been possible without the assistance of Marjan Bolle, our conference secretary.

Wiet Bouma and Hugo Velthuijsen
PTT Research
St. Paulusstraat 4
2264 XZ Leidschendam
The Netherlands
Tel: +31 70 332 5457/6258
Fax: +31 70 332 6477
Email: {L.G.Bouma, H.Velthuijsen}@research.ptt.nl

Leidschendam, March 1994

References

[1] T.F. Bowen, F.S. Dworak, C.-H. Chow, N.D. Griffeth, G.E. Herman, and Y.-J. Lin. Views on the feature interaction problem. In *Proceedings Seventh International Conference on Software Engineering for Telecommunications Switching Systems*, pages 59–62, Bournemouth, July 1989.

[2] E.J. Cameron, N.D. Griffeth, Y.-J. Lin, M.E. Nilson, W.K. Schnure, and H. Velthuijsen. A feature interaction benchmark for IN and beyond. In L.G. Bouma and H. Velthuijsen, editors, *Feature Interactions in Telecommunications Systems*, pages 1–23, Amsterdam, May 1994. IOS Press.

[3] E.J. Cameron and Y.-J. Lin. A real-time transition model for analyzing behavioral compatibility of telecommunications services. In *Proceedings ACM SIGSOFT '91 Conference on Software for Critical Systems*, pages 101–111. ACM Press, December 1991.

[4] E.J. Cameron and H. Velthuijsen. Feature interactions in telecommunications systems. *IEEE Communications Magazine*, 31(8):18–23, August 1993.

[5] J.Y. Halpern and Y. Moses. Knowledge and common knowledge in a distributed environment. *Journal of the ACM*, 37(3):549–587, July 1990.

[6] Y. Inoue, K. Takami, and T. Ohta. Method for supporting detection and elimination of feature interaction in a telecommunication system. In *Proceedings International Workshop on Feature Interactions in Telecommunications Software Systems*, pages 61–81, St. Petersburg, FL, 2-4 December 1992.

A Feature Interaction Benchmark for IN and Beyond

E. Jane CAMERON, Nancy D. GRIFFETH, Yow-Jian LIN
Margaret E. NILSON, William K. SCHNURE
Bellcore, 445 South Street, Morristown, NJ 07962-1910, USA

Hugo VELTHUIJSEN
PTT Research, St. Paulusstraat 4, 2264 XZ Leidschendam, The Netherlands

Abstract. Rapid creation of new services for telecommunications systems is hindered by the *feature interaction* problem. This is an important issue for development of IN services, not only because of interactions among IN services themselves but because of interactions of IN services with switch-based services and potential interactions with services not yet developed. Furthermore, the problem is fundamental to service creation; it is not restricted to IN services. Any platform for telecommunication services requires a method for dealing with the feature interaction problem. A number of approaches for managing feature interactions have been proposed. However, lack of structured ways to categorize feature interactions makes it difficult to determine if a particular approach has addressed some, if not all, classes of interactions.

We describe and analyze a number of feature interactions by using two independent classification schemes. This paper is a step to achieving the goal of a coherent industry-wide collection of illustrative features and their interactions. The collection will help convey the scope of the feature interaction problem. It will also serve as a benchmark for determining the coverage of various approaches, and as a guideline for identifying potential interactions in software architectures and platforms.

1 Introduction

This paper outlines two methods for categorizing feature interactions. The goal is to improve understanding of the scope of the feature interaction problem in telecommunications systems and to analyze the coverage of approaches to solving the problem.

Features in a telecommunications system are packages of incrementally added functionality providing services to subscribers or the telephone administration [7]. For the purposes of discussion in this paper, the meaning of feature will be interpreted broadly. Some features, such as *Plain Old Telephone Service (POTS)*, offer basic means of setting up conversations and billing for services,[1] whereas others, such as *Call Forwarding*

---

[1] We consider POTS as a feature, so that we can address all interactions in a uniform way even when POTS is involved.

or *Call Waiting*, modify the basic telephone service that is offered to subscribers. Still others, such as those features and services offered by *ISDN* rely on sophisticated transport protocols to carry signals to and from subscribers, thus increasing the signaling capabilities and modifying the software protocols for controlling calls. In this paper, we will not make a distinction between the terms *feature* and *service*.

Feature interactions are understood to be all interactions that interfere with the desired operation of the feature and that occur between a feature and its environment, including other features or other instances of the same feature. Additionally, interference of one part of a feature with another part of that feature (e.g., in case of a distributed implementation of a feature) is considered to be a feature interaction.

In order to add a new feature to a telecommunications system, its interactions with the telecommunications system and other existing features must be managed in some fashion. A number of approaches have been proposed to address the feature interaction problem [2, 4, 6, 8, 9, 10]. While the increasing attention paid to this problem is encouraging, many approaches that claim to have addressed the problem fully in fact only solve a part of it. This is not surprising, since the problem itself is often vaguely defined through the use of instances of potential interactions or definition is avoided altogether. There has been no coherent collection of feature interactions available to illustrate the full range of the problem. This paper provides such a collection and proposes a taxonomy for categorizing it. This taxonomy is intended to serve as a benchmark for determining the coverage of various approaches.

The problem of feature interactions has created difficulties in the process of service deployment. Recent and foreseeable changes in the telecommunications industry complicate the problem further. Important changes include:

- the process of service creation is no longer largely governed by a single organization as it was prior to the divestiture of Bell System;

- the platform for an intelligent network promotes independent and rapid deployment of customized features by operating companies and their associated suppliers and independent information providers;

- the telecommunications network is becoming increasingly heterogeneous, with equipment and support software provided by several suppliers; and

- the service logic for controlling call processing is becoming increasingly distributed, with service logic programs distributed among various network components.

When a large number of features and equipment deployed by numerous suppliers must work together, the traditional approach of a manual feature-by-feature analysis is no longer feasible. Clearly, more powerful techniques for the management of feature interactions are needed.

In this paper, we present two different ways of categorizing feature interactions: by the *nature* of the interaction and by the *cause* of the interaction. The categorization by the *nature* of the interaction is defined by the following three dimensions: the *kind of features* involved; the *number of users* involved; and the *number of network components* involved. The categorization by the *cause* of the interaction includes: violation of assumptions about the system; limitations on network support; and problems intrinsic to large, heterogeneous distributed systems. An illustrative collection of feature interactions is presented, categorized by the two classification schemes, and cross-referenced.

However, interactions involving system security, mixed media, and administrative features are not discussed.[2]

We find that our taxonomy is illuminating when attempting to analyze various proposals that address the feature interaction problem. By identifying and classifying various issues regarding the feature interaction problem, we are now able to port ideas from different areas of computer science to address the appropriate issues (see also Sections 4 and 5). In addition, we believe that this benchmark can also serve as guidance for the development of more powerful service creation tools and environments. Note that the feature interaction problem is interpreted more broadly in this paper than is usually done elsewhere. Traditionally, the feature interaction problem only refers to situations where proper interworking of a group of features is not possible. Here we also include situations where the resulting joint behavior is not what the user would reasonably expect from understanding each individual feature alone. Such situations could arise because some feature descriptions are incomplete or ambiguous, or because the user does not envision all the implications of activating one feature, or because the user is not aware of some subtle differences in telecommunication terminology (e.g., that the directory number (DN) and the dialed number are not necessary the same).

In order to keep the examples simple and focused on the problem, we also elected to use features that are widely deployed (hence familiar to many telephone users) for illustrating feature interactions. We will briefly describe the features when they are used. The names and descriptions of these features are based on the collection of feature definitions in [1]. Some examples presented in this paper have actually been resolved in the current public network; therefore, they may not appear to be instances of the problem to some users. Nevertheless, it is worthwhile to recognize them as interactions, since the interactions have been handled case-by-case, without using or producing a global set of guidelines for handling interactions.

The paper is organized as follows. Section 2 presents a categorization of feature interactions by their *nature*. Also, all of our prototypical examples are introduced in this section. Section 3 provides a categorization of feature interactions by their *cause(s)*, referring back to the examples of Section 2. Section 4 summarizes our categorizations and indentifies potential approaches in managing feature interactions. The paper ends with some concluding remarks in Section 5.

2 Categorization of the Nature of Interactions

This categorization has three dimensions: the kind of features involved in the interaction, the number of users involved in the interaction, and the number of network components in the interaction.

The first dimension distinguishes interactions that involve only *customer features* from interactions that involve *system features*, instead of or in addition to customer features. Customer features include all of the call processing features visible to the general public, such as *Call Waiting, Call Forwarding, 800 services*, features offered by *CENTREX*, and so on. System features include billing and other Operations, Administration, and Maintenance (OA&M) services. Interactions can occur among customer features; among customer features and system features; or among system features.

---

[2]These categories probably hold for these kinds of features as well as they do for the features presented here. Presenting a truly comprehensive set of features would take far too much space, without adding significantly to the substance of the benchmark.

The second dimension distinguishes *single-user* interactions from *multiple-user* interactions. Single-user interactions arise when different features, while simultaneously activated by a single user, interfere with each other, whereas multiple-user interactions arise when features activated by one user interfere with those activated by another user. The user of a feature is the one whose call processing logic includes the functionality of the feature; the user of a feature could be different from the subscriber to a feature, i.e., the one who pays for the feature. As an example, the municipal government that provides *911* emergency services to its community is the subscriber of *911*, but the user of *911* is the one who makes a *911* call: the logic of *911* is incorporated into the call processing logic of every line in the municipality. As another example, the user of the residential *Call Forwarding* feature can only be its subscriber: the logic of *Call Forwarding* is part of the call termination treatment associated with the subscriber only.

The third dimension distinguishes between *single-component* interactions, which arise when only one network component (switching system, adjunct, SCP, services node, CPE) is involved in the processing, and *multiple-component* interactions, which arise when features supported on one network component interfere with the operation of features supported on another network component. Multiple-component interactions are becoming more common as features are increasingly being supported by a distributed architecture. These interactions are especially difficult to resolve, since they may require distributed control or inter-supplier collaboration to get features to interoperate correctly.

In this section, the examples of feature interactions are presented in the following order:

- SUSC (Single-User-Single-Component) interactions among customer features;
- SUMC (Single-User-Multiple-Component) interactions among customer features;
- MUSC (Multiple-User-Single-Component) interactions among customer features;
- MUMC (Multiple-User-Multiple-Component) interactions among customer features;
- CUSY (CUstomer-SYstem) interactions among customer features and system features.

Since we found many more examples of interactions among customer features, we focus in this paper mainly on interactions that interfere with the expected functioning of customer features. The full range of choices among the number of users and number of components is addressed only for interactions among customer features. Some interactions among customer features and systems features (which are mostly multi-component interactions) are also discussed. Interactions that involve only system features are not addressed at all in this paper.

Some multiple-user interactions can be either MUSC type or MUMC type, depending upon whether all users involved are associated with a single network component. In our categorization we put all such multiple-user interactions in the MUMC category; only those multiple-user interactions that arise specifically in a single network component are categorized as MUSC interactions. By the same token, some single-user interactions can be either SUSC type or SUMC type, depending upon whether all features involved are deployed in a single network component or distributed among several. In this case, we put all such single-user interactions in the SUSC category; only those single-user interactions that explicitly involve multiple network components are put in the SUMC category. Of course, with a distributed architecture such as that of AIN, many SUSC interactions could also have been defined as SUMC interactions.

In most cases, we will give several examples for the same category of feature interactions (according to the present categorization). In Section 3, these examples will be used to illustrate a further categorization of feature interactions by their causes.

2.1 SUSC Interactions

SUSC (Single-User-Single-Component) interactions occur because incompatible features are simultaneously in use by a single user in a single network component such as a switching element or a service control point (SCP). Some of these interactions arise because of *functional ambiguities* between the features (that is, two different features are designed to deal with the same call processing situation, but differ as to how it will be handled), or because of *interferences* (that is, one feature will preclude the proper execution of another feature). Other of these interactions arise because of resource limitations or signaling limitations.

There are many combinations of features that produce functional ambiguities. Features such as *Call Waiting*, *Call Forwarding*, *Answer Call* (or *Voice Mail*), and *Automatic Recall* provide alternatives that enable the calling party to contact the called party on calls that would otherwise be unsuccessful. A person who subscribes to more than one such feature could encounter interactions when features compete for call control.

Example 1 *Call Waiting* and *Answer Call*.
When a call attempts to reach a busy line, *Call Waiting* generates a call-waiting tone to alert the called party, whereas *Answer Call* connects the calling party to an answering service. Suppose that **A** is a subscriber of both features. If **A** is already on the line when the second call comes in, should **A** receive a call-waiting tone or should the second call be directed to the answering service?
□

Of course, these interactions can be resolved simply by establishing precedence relations among conflicting features. The resolution can be either determined for all subscribers during service creation or personalized during provisioning. Note that this is not to be confused with "signal ambiguities" (cf. Example 2) which can be resolved by a richer set of functional signals.

Example 2 *Call Waiting* and *Three-Way Calling*.
The signaling capability of customer premise equipment (CPE) is limited. As a result, the same signal can mean different things depending on which feature is anticipated. For example, a flash-hook signal (generated by hanging up briefly or depressing a 'tap' button) issued by a busy party could mean to start adding a third party to an established call (*Three-Way Calling*) or to accept a connection attempt from a new caller while putting the current conversation on hold (*Call Waiting*). Suppose that during a phone conversation between **A** and **B**, an incoming call from **C** has arrived at the switching element for **A**'s line and triggered the *Call Waiting* feature that **A** subscribes to. However, before being alerted by the call-waiting tone, **A** has flashed the hook, intending to initiate a three-way call. Should the flash-hook be considered the response for *Call Waiting*, or an initiation signal for *Three-Way Calling*?
□

A similar situation occurs when lifting a handset is interpreted as accepting the incoming call, even though the user's intention in doing so is to initiate a call—remember the cases when one picks up the phone in the absence of ringing and somebody is already on the other end of the line. The call processing is behaving just as it was designed to, but some users may be momentarily puzzled.

Call Waiting and *Three-Way Calling* can also interact due to resource limitations: since both features need a three-way bridge, according to [1] one feature is automatically disabled once the bridge is used by the other. Nevertheless, as a later example (cf. Example cwtwc2) will show, even when the simultaneous use of both features are possible, more confusion could be introduced.

Now we present some interference examples where one feature will not function properly in the presence of another feature.

Example 3 *911* and *Three-Way Calling.*
A *Three-Way Calling* subscriber must put the second party on hold before bringing a third party into the conversation. However, the *911* feature prevents anyone from putting a 911 operator on hold. Suppose that **A** wishes to aid a distressed friend **B** by connecting **B** to a *911* operator using the *Three-Way Calling* service. If **A** calls **B** first and then calls 911, **A** can establish the three-way call, since **A** still has the control of putting **B** on hold before calling 911. However, if **A** calls 911 first, then **A** cannot put the *911* operator on hold to call **B**; therefore **A** cannot make the three-way call.

Assume that **A** has made the three-way call according to the former scenario. Whether the *911* operator has the control over the entire call remains ambiguous: it appears that the *911* operator does not have control over **B**, even though the operator has control over **A**. Therefore, the operator cannot prevent **B** from dropping out of the call.
□

The main reason for preventing a caller from dropping out of a 911 call is to facilitate tracing the origin of the call. Since the ID of a caller can be delivered in an advanced network, the "no interruption" policy of *Emergency 911* could be replaced by a mandatory *Calling Number Delivery*. Consequently, the above-mentioned interaction regarding setting up a three-way call with a 911 operator can be resolved. On the other hand, which ID gets delivered becomes a new issue. For example, can the *911* operator see the ID of **B**, in addition to that of **A** who sets up the three-way call?

Example 4 *Terminating Call Screening* and *Automatic ReCall.*
Terminating Call Screening assumes that every incoming call will be screened against the incoming call screening list. However, *Automatic ReCall* (i.e., automatically returning the last incoming call), if processed before *Terminating Call Screening* when the line is busy, could just register the incoming call number without having the number screened. Suppose that **A** has both features, and **B**'s number is among those that **A** refuses to accept via the *Terminating Call Screening* feature. Now, if **B** calls **A** while **A**'s line is busy, then when **A**'s line becomes idle the *Automatic ReCall* feature will initiate a connection attempt back to **B**, which nullifies the purpose of the *Terminating Call Screening* feature.
□

The AIN framework [4] requires local switching elements to trigger service logic programs stored in SCPs, adjuncts, or other network components. Restrictions imposed by suppliers on the triggering capabilities of some switching elements can cause interactions, as we will show in the next example.

Example 5 *Originating Call Screening* and *Area Number Calling*.
Originating Call Screening aborts attempts to connect the subscriber to directory numbers in the screening list, whereas *Area Number Calling* decides the actual terminating number based upon the originating number and the dialed number. Each of them needs to launch a query for call origination treatment during call set up. Suppose that **A**, who is a subscriber of the *AIN Originating Call Screening* feature, calls Domino's area number to order a pizza. If the switching element imposes a restriction on the number of queries per call, e.g., one query per call, then after launching a query for the *Originating Call Screening* feature that **A** subscribes to, the component will not be able to launch another query for the *Area Number Calling* feature that Domino's Pizza pays for to serve its customers.
□

The reason for having an upper bound on the number of queries per call is to ensure that call origination/termination treatment will not go into an infinite loop. However, just how many queries per call is sufficient is unclear. Two queries per call may be enough for the above-mentioned situation, but is too low if **A** also has an AIN *Speed Calling* feature that triggers another query to get the dialed number translated.

2.2 SUMC Interactions

SUMC (Single-User-Multiple-Component) interactions form an increasingly important problem of coordination in a telephone network where features accessible to one customer are deployed in different network components. Making sure that features residing in different network components behave correctly is not straightforward; careful consideration must be given to the information that should be shared among the features, and to the order of execution. An interaction arises when the existence of one feature is not known to or has not been considered by designers of features in other network components, resulting in the loss of some feature functionality. The following is one such interaction.

Example 6 *Operator Services* and *Originating Call Screening*.
Operator Services may be handled in a remote switching element that does not have access to the feature subscription profile of every customer who wishes to use the services. Therefore, every call made through *Operator Services* acts like an outgoing POTS call, except that it is operator-assisted. Suppose that **A** has subscribed to the *Originating Call Screening* feature deployed in a local switching element, hoping to screen outgoing calls made to any number in a screening list. However, since the remote switching element does not have the screening list, it is possible for anybody to make an operator-assisted 0+ or 0– call to a screened number by **A**'s line.
□

Timing is also critical to the correct inter-working of features deployed in multiple network components. During call processing, the thread of control[3] can switch from a

---

[3] We assume that each feature defines a computation thread, and all activated features constitute a group of concurrent threads. The thread of control is in one feature if the system is executing the code corresponding to that feature.

feature controlling one stage of call processing to another feature controlling another stage. When two features are deployed in different network components the transition time from one thread to another may be significant. However, the transfer of control thread may be hidden from the customer. In many cases a customer may send a signal (an off-hook, a "#", or some digit) too early or too late; consequently, the signal is not interpreted by the feature the customer expected. Here is a situation that illustrates this point.

Example 7 *Credit-Card Calling* and *Voice-Mail service.*
Instead of hanging up and then dialing the long distance access code again, many credit-card calling services instruct callers to press [#] for placing another credit-card call. On the other hand, to access voice mail messages from phones other than his/her own, a subscriber of some *Voice-Mail service* such as *Aspen* can (1) dial the Aspen service number, (2) listen to introductory prompt (instruction), (3) press [#] followed by the mailbox number and passcode to indicate that the caller is a subscriber, and then (4) proceed to check messages. However, when a customer places a credit-card call to Aspen, the customer does not know exactly when the *Credit-Card Calling* feature starts passing signals to Aspen instead of interpreting them itself. Suppose that **A** has frequently called Aspen and knows how to interact with Aspen. When **A** places a credit-card call to Aspen, **A** may hit [#] immediately without waiting for the Aspen's introductory prompt. However, the [#] signal could be intercepted by the credit-card call feature; hence it is interpreted as an attempt to make a second call.
□

Timing is not the only factor that contributes to the above problem; limited signaling capability causes the two features to choose the same signal for customer communication. Two obvious solutions present themselves: synchronizing call processing with the user to avoid timing problems, or expanding the signaling capabilities to disambiguate the user's intention. However, even when the signal set is expanded, if features are supplied by many independent suppliers, it will be difficult to resolve timing problems or to make sure that signals are used unambiguously.

Because of the trend toward a distributed architecture for controlling call processing, the number of possible interactions in this category is increasing. One such trend is the deployment of AIN features in SCPs and adjuncts, separate from the switch-based features that reside in local switching elements. Currently, local switching elements serve as the front-end for the subscribers to the telecommunications network. Unless a local switching element can coordinate the execution of any possible subset of features subscribed to by a user,[4] regardless of their distribution, many more SUMC-type feature interactions will occur between local switching elements and SCPs or adjuncts. The following interaction between an AIN feature and a non-AIN feature is related to how a local switching element chooses to query an SCP or an adjunct.

Example 8 *Multi-location Business Service-Extension Dialing (MBS-ED)* and *CENTREX*
The *AIN Release 0 MBS-ED* feature allows a customer to extend a 4 or 5 digit extension dialing plan to locations served by different switching elements. This is accomplished by querying an SCP or an adjunct to translate digit combinations to directory numbers. *CENTREX* features are also based upon a 4 digit extension dialing plan, but are

---

[4]Other possibilities are to have the coordination done at some other network components (e.g., SCPs), or to implement some distributed coordination mechanism across the network.

served by a single switch. Thus, *CENTREX* features do not query an SCP or adjunct, as the corresponding directory numbers must be local to that switch. When a 4-digit number is received in a switch that supports both *AIN Release 0 MBS-ED* and *CENTREX* features, it is not clear whether an SCP should be queried or not. Assignment of disjoint subsets of the 4-digit numbers to the *MBS-ED* features and the *CENTREX* features could be used by the switch to detect what type of feature is in effect whenever a 4-digit number is received.
□

2.3 MUSC Interactions

MUSC (Multiple-User-Single-Component) interactions can occur when two or more customers access the features associated with a physical line. This type of problem can usually be avoided if the group of customers that share the features carefully work out a consistent use of features. But as the network architecture evolves, new kinds of MUSC interactions will arise. Personal Communication Services (PCS) [5], while providing mobility based on dynamic binding[5] of subscribers to CPEs, introduce a new set of MUSC interactions when features of multiple customers compete for the same physical resources. Examples in this section illustrate MUSC interactions in the present and in foreseeable telecommunication networks.

Example 9 *Call Forwarding* and *Originating Call Screening.*
Call Forwarding allows incoming calls to be redirected to another directory number, while *Originating Call Screening* aborts attempts to connect the subscriber to some other directory numbers. It may seem unlikely that anyone would intentionally block calls to a directory number and then forward calls to the same number, but this situation can arise if the forwarding number and the call screening number were supplied by two different people sharing the same physical line. For example, if an adolescent knows that his parents have used *Originating Call Screening* to block all calls to a dial-porno number **X** from their line **A**, he may instruct the switching element to forward all calls terminated at **A** to **X**, and then call himself to get the effect of *Call Forwarding*. Subscribers of *Originating Call Screening* will not be satisfied if a loophole like that exists. Whether the forwarding number is considered a dialed number (to be checked against the screening list) becomes an issue. Supposing that *Call Forwarding* takes precedence over *Originating Call Screening*, calls can be forwarded to the forwarding number despite the fact that the number is also on the screening list.
□

Personal Communication Services are attractive, but the mobility and convenience of PCS also introduce new kinds of MUSC interactions. For one thing, since every customer can choose to associate services with any CPE, resource contention is a potential problem. There could be many more interactions to be resolved. For example, when multiple PCS customers are currently associated with a single CPE, some switch-based features may not work properly based upon the status of the CPE (or the physical line connected to the CPE) alone, as we show in the following.

---

[5]Part of setting up mobile services is to determine the location of the subscriber/user. Hence, the particular CPE that will receive this subscriber's calls must be determined on a per call basis. This is different from the more traditional static binding, where the location of the CPE designated to receive calls for a given subscriber/user is stored in a table. Once the location is stored, it will be used for all calls, until it is replaced.

Example 10 *Call Waiting* and *Personal Communication Services*
Call Waiting is a feature assigned to a *directory number*. However, *Call Waiting* uses the status of the *line* with which the number is associated to determine whether the feature should be activated: at present in a public switched telephone network, if a non-ISDN line is in use, then it is busy; a second call to the same line will trigger the switching element to send out a call-waiting tone. PCS customers may not all be subscribers of *Call Waiting*. Suppose that **X** and **Y** are both PCS customers currently registered with the same CPE; **X** has *Call Waiting* but **Y** does not. We further assume that **Y** is on the phone when somebody calls **X**. Since **X** has *Call Waiting* and the line is busy, the new call triggers the *Call Waiting* feature of **X**. But is it legitimate to send the call-waiting alert through the line to interrupt **Y**'s call? If not, then **X**'s *Call Waiting* feature is ignored.
□

2.4 MUMC Interactions

MUMC (Multiple-User-Multiple-Component) interactions can occur when two or more users access features supported on multiple network components.

Consider that POTS is a basic communication protocol for telephone users making phone calls, and additional features subscribed to by each user vary the POTS protocol a bit. Since two customers can subscribe to different sets of features, potentially any customer may need to communicate with another customer whose protocol is somewhat different. Identifying and classifying these variations is an important step to designing a software infrastructure that enables these more complex variations of the basic protocol to correctly interoperate. The types of interactions that can arise depend on how a feature alters the logic of call control, or the environment or assumptions needed for another feature to function as expected. Here we begin with some examples that are caused by conflicting assumptions about the use of dialed numbers, directory numbers, or physical lines as the identifier of customers.

Example 11 *Originating Call Screening* and *Multiple Directory Number Line with Distinctive Ringing.*
Originating Call Screening is a feature based on directory numbers; any call placed to a directory number on the screened list will be blocked. Disallowing calls to a directory number prevents connections to the identified line, provided the line is associated with only that directory number. However, services for *Multiple Directory Number Line with Distinctive Ringing (MDNL-DR)* allow more than one directory number to be associated with a single line. Suppose that **B** is a subscriber of the *MDNL-DR* service with two numbers **X** and **Y**, and **A** has the *Originating Call Screening* service with the number **X** in the outgoing call screening list. **A** can still make calls to **B**'s line, if **A** dials **Y**.
□

It can be argued that *Originating Call Screening* is doing what it was designed to, i.e., screening only directory numbers in the list. Therefore, the above example is not a feature interaction. The question raised here, though, is whether customers may misinterpret the intention of the feature when they subscribe to it. Surely, in the above example **B** could have two physical lines instead of *MDNL-DR*, and the scenario will be the same. To avoid any connection to **B**, both directory numbers must be in the screening list. However, in practice a user of *Originating Call Screening* should not have the false expectation that the feature can be used to block any connection to a particular person, because the feature can only screen based on directory numbers; said person may answer calls on some other "unblocked" line, especially if this person is a PCS user.

Example 12 *Originating Call Screening* and *Call Forwarding. (revisited)*
Originating Call Screening blocks calls based on the number dialed; thus, calls to a particular line are blocked only if the dialed number is associated with that line. However, *Call Forwarding* connects to a line other than the one associated with the dialed number. If **B** forwards all incoming calls to the number **X**, connections from **A** to the line identified by **X** will be established, when **A** calls **B**, even if **A** has **X** on the outgoing call screening list.
□

In a distributed environment such as a telecommunications system, lack of a consistent global view of call control among the parties involved in a call often causes interactions as well. The inconsistency may be due to inability to convey the line status, as the next example illustrates; or it may be caused by competing control relationships, as the ones to be discussed in Example 14 and 15.

Example 13 *Call Waiting* and *Automatic CallBack.*
Automatic CallBack is triggered if the called party is busy. But a line with *Call Waiting* appears to be idle to a caller, although it is actually busy. Suppose that **B** is a *Call Waiting* subscriber and **A** has activated the *Automatic CallBack* feature. Will *Automatic CallBack* work for **A**, if **B** is talking to **C** when **A** calls **B**? Note that because of the *Call Waiting* feature that **B** has, **A** will receive a ring-back signal from **B**.
□

Even if a feature is working correctly, confusing situations can still occur when the feature is invoked by both parties to a call. The example below shows how a confusing situation can arise. When two people use the *Call Waiting* feature, complicated call control relationships arise. Although behaviors of the three parties involved in a single *Call Waiting* scenario are specified, nothing is mentioned about the many possible combined scenarios when a chain of people use *Call Waiting* concurrently—a typical case of incomplete specification.

Example 14 *Call Waiting* and *Call Waiting.*
Call Waiting allows a subscriber to put the other party on hold. However, it does not protect the subscriber from being put on hold. Confusion can arise when two parties exercise this type of control concurrently. Suppose that both **A** and **B** have *Call Waiting*, and **A** has put **B** on hold to talk to **C**. While on hold, **B** decides to flash the hook to answer an incoming call from **D**, which puts **A** on hold as well. If **A** then flashes the hook expecting to get back to the conversation with **B**, **A** will be on hold instead, unless either **B** also flashes the hook to return to a conversation with **A** or **D** hangs up automatically returning **B** to a conversation with **A**.

An ambiguous situation arises, when **B** hangs up on the conversation with **D** while **A** is still talking to **C**; there are two separate contexts in which to interpret **B**'s action. Assume that *CW1* refers to the *Call Waiting* call among **C-A-B** and *CW2* refers to the one among **A-B-D**. According to the specification of *Call Waiting*, in the context

of *CW2* **B** will be rung back (because **A** is still on hold) and, upon answering, become the held party in the *CW1* context and hear nothing. But, in the context of *CW1* the termination **B** will be interpreted as simply a disconnection, thus **A** and **C** are placed in a normal two-way conversation, and **B** is idled. The question is: Should **B** be rung back or should **B** be idled?
□

Removing signal ambiguities and contention for resources may initially seem like plausible solutions for solving interactions like that between *Call Waiting* and *Three-Way Calling* (cf. Example 2). However, because of the more complex call control, the possibility would now exist that each party to a call may be involved in two conversations and may be alternating between them. Truly complex situations could arise, as is shown below.

Example 15 *Call Waiting* and *Three-Way Calling (revisited)*.
Consider how *Call Waiting* and *Three-Way Calling* might interact in the situations where a user can exercise both features imultaneously on the same line.[6] The call control relationship can now become quite complicated. Suppose that **A** has both *Call Waiting* and *Three-Way Calling*, **B** has *Call Waiting*, and **A** is talking to **B**. Now **C** calls **A**, so **A** uses *Call Waiting* to put **B** on hold and talks to **C**. **A** may decide to have **B** join his conversation with **C**, so he puts **C** on hold, makes a second call to **B**, and after **B** answers the call with *Call Waiting*, **A** brings **C** back into the conversation to establish a three-way call. There are three contexts in this establishment: a *Call Waiting* call and a *Three-Way Calling* call, both established by **A** among **B-A-C**, and a *Call Waiting* call established by **B** as **A-B-A**. Now, if **B** hangs up, then according to the contexts established by **A**, the session becomes a two-way call between **A** and **C**; according to the contexts established by **B** though, **B** should get a ring-back because **B** still has **A** on hold.
□

The goals of different features may be in direct conflict with one another. Such conflicts prevent certain combinations of features from being used simultaneously, unless one is assigned priority over another. Next is a simple example of one such conflict, that is presently resolved assigning one feature priority over the other.

Example 16 *Calling Number Delivery* and *Unlisted Number*.
Calling Number Delivery is a call-processing feature that delivers the directory number of the calling party to the customer's premises during the ringing cycle; this assumes that information such as the subscriber's number will be released. *Unlisted Number*, on the other hand, is a directory-service feature designed to allow a subscriber to keep the number private. Conflicts between the goals of these two features arise when a customer **A** with *Unlisted Number* places a call to another customer **B** with *Calling Number Delivery*. If the network allows **A**'s number to be delivered to **B**, then **A** loses privacy; if it doesn't, then **B** gets no information. Either way, one of the features doesn't perform its intended function. Currently, delivery of the number can be blocked by the *Calling Number Delivery Blocking* call-processing feature, which nullifies the *Calling*

---

[6]The current feature requirements state that *Call Waiting* and *Three-Way Calling* cannot be activated simultaneously on the same line. However, in some switching implementations it is possible for a user, who has set up a three-way call with *Three-Way Calling*, to put the other two parties on hold to answer another incoming call with *Call Waiting*.

Number Delivery feature by delivering a number consisting of only 1's.
□

Many features are equipped with parameters that can be (or need to be) instantiated (assigned values). For example, to use *Call Forwarding*, a subscriber must supply the forwarding number. The following simple example illustrates that this can lead to numerous *livelock* situations.

Example 17 *Call Forwarding* and *Call Forwarding*.
A customer with *Call Forwarding* can redirect calls to any number. As a result, any one can "accidentally" forward all incoming calls to his/her own number and create an infinite loop. Moreover, calls for **A** can be forwarded to **B**, only to be forwarded back to **A** by **B** (i.e., a two-number loop); or to **B**, then to **C**, then back to **A** (i.e., a three-number loop); and so on. Detecting the loop during call processing is difficult, if the amount of information (e.g., the numbers having been reached) passed is limited. Currently, SS7 uses a counter, aborting calls that are forwarded more than 5 times.
□

The following example shows how timing can cause interactions in telephone service logic programs, just as it does in many distributed software systems. The following example was first mentioned in [7].

Example 18 *Automatic CallBack* and *Automatic ReCall*.
When a subscriber dials the number X of a line that turns out to be busy, the feature *Automatic CallBack* (AC) can be activated. *Automatic CallBack* automatically redials **X**, when X becomes idle. If **X** is idle, the call is considered completed, and the AC request is removed immediately. On the other hand, if **X** is busy, the AC request is placed on a queue of AC requests in the central office. Call set-up will be attempted again when **X** becomes idle, the subscriber's line is idle, and the subscriber has answered a special ringback. On the other hand, *Automatic ReCall* (AR) automatically returns the latest incoming call for the subscriber, whether the call from some line **Y** was answered or not. If **Y** is idle, the call completes and the AR request is removed. If **Y** is busy, the AR request is placed on a queue of AR requests in the central office. Call set-up will be attempted again when **Y** becomes idle, the subscriber's line is idle, and the subscriber has answered a special ringback, very much like the way *Automatic CallBack* operates. Suppose that **A** has the *Automatic CallBack* feature and **B** has the *Automatic ReCall*. If **B**'s line is busy when **A** calls **B**, it is possible that thereafter the switching element serving **A**'s line keeps initiating new calls to **B** while the switching element serving **B**'s line stays busy returning every call **A**'s AC feature made. Currently, no provisions are made to disallow such behavior based on the assumption that synchronicity of the execution of the two features is highly unlikely. If necessary, the features could be implemented, using an artificial time delay, in such a way that synchronous execution is impossible.
□

2.5 CUSY Interactions

The CUSY (CUstomer-SYstem) category of feature interactions refers to interactions between a customer feature and any system feature for operations, administrative services, or maintenance. One interesting area is the interactions of some customer features with the billing system. Since the divestiture of the Bell system, a simple phone call may now involve more than one service provider or administrative domain. There is also a trend to bill an increasing number of features on a per usage basis. However, it is a general policy that the usage of a specific feature should not be billed, unless the call that uses the feature is completed. This simple view creates confusion, because the definition of a "complete call" becomes ambiguous in multiple administrative domains. The basic question is: if one service provider has completed its part of service for a call but another one did not, does anyone get paid?

Example 19 *Long distance calls* and *Message Rate Charge services.*
Each long distance call consists of at least three segments—two local accesses at each end and one provided by an interexchange carrier in between. Should a customer be charged for the segments that have been successfully completed even if the call did not reach its final destination? Would it be counted as one unit toward the total local units allowed per month for a *Message Rate Charge service*?
□

Some "add-on" billing policies, such as phone access charges introduced by third-party service providers for hotels and resorts, present yet another view of "call completion". If customers assume that they will be billed for completed calls, but not for incomplete calls, then confusion can arise if a complete call is not established.

Example 20 *Calling from hotel rooms.*
Many hotels contract with independent vendors to collect access charges for calls originated from phones in their premises. Without being able to access to the status of call connections, some billing applications developed by these vendors use a fixed amount of time to determine if a call is complete or not—thus one can be billed for incomplete calls that rang a long time, or not billed for very short duration calls (even long distance).
□

When multiple services in AIN Release 0 are invoked on one call, the interface between the SSP (Service Switching Point) and the SCP may not be adequate should services be billed on a per usage basis. The following example illustrates the situation.

Example 21 *Billing in AIN Release 0.*
In AIN Release 0, the SCP instructs the SSP to generate a billing (AMA) record by sending a call-type code (3 numerics) to the switching element. Because there is room for only one call-type code in the TCAP[7] response, a problem arises when multiple services are invoked by one SCP query. For example, if Area Number Calling is assigned a call-type code of 271 and Originating Call Screening is assigned a call-type code of 275, how will the SCP tell the SSP to generate an AMA record that reflects the fact of both features being used? Of course, one solution would be to assign another call-type code to indicate that both features were invoked. However, with the proliferation of AIN services, call-type codes will soon be exhausted[8].
□

---

[7]TCAP (*Transaction Capabilities Application Part*) is a protocol in the Common Channel Signaling 7 (CCS7) for a Service Switching Point (SSP) to communicate with a Service Control Point (SCP) in AIN.

[8]Note that this approach would need 2^{10} different call-type codes if only 10 features were deployed, exceeding the 1000 available 3-numerics codes

In an AIN architecture, when a subscriber of any AIN-based service tries to make a call, it is necessary to consult the record with information for that customer in the SCP/adjunct database. Since the provisioning of AIN-based services now involves more than one network component, the way provisioning is carried out is significant.

Example 22 *AIN-based services* and *POTS*.
In the provisioning of some *AIN-based services* deployed in SCPs, updates must normally be made to two network components: the SSP must now know that a trigger should be established to query the SCP databases; and the SCP should have an updated record of customer information. Let **A** be a new subscriber of the *AIN-based services*. Suppose that the provisioning for **A** is in progress, and that (1) once the trigger is set, the SSP will query the SCP database on all calls; (2) the trigger is set for **A** but the customer record is not updated yet in SCP; and (3) **A** has started making a POTS call. In this scenario, the query triggered by **A**'s POTS call will result in a "customer record not found" response from the SCP; consequently, even the POTS call cannot go through.
□

3 Categorization of Causes of Interactions

The examples presented in Section 2 suggest that there are many different causes of feature interactions. These include violation of feature assumptions, limited network support, and problems in distributed systems in general. In the following we address these causes in detail.

3.1 Violation of Feature Assumptions

Features in a telecommunications network need to operate under a set of assumptions. These assumptions may include a particular call processing model, some architectural support, a special billing system, or even how customers perceive the network operates. In a long-lived system such as telecommunications software, the evolution of system architecture or additions to its feature set often create a new environment that violates the assumptions of existing features. We identify several kinds of assumptions that, once violated, can result in some interactions.

3.1.1 Assumptions about Naming

Historically, a telephone number uniquely identified a telephone line. Thus, anyone who placed a call to a number, no matter where the call originated, reached the same telephone line every time. Also, individual customers who were associated with a specific line (e.g., members of the same household) formed a single entity and did not need be further distinguished. New features enabling rerouting or providing personal identification for multiple persons assigned to a specific line have made this "one_number-one_line-one_entity" model obsolete. Still, subscribers who are used to this model may be confused by its violation.

Features such as *800-services*, *900-services*, *911*, and *411* enable calls to be routed based on parameters such as the calling party's location, the time of day, or the day of the week. They alter the one_number-one_line assumption, but still honor the one-number-one-entity view, for entities like emergency unit (*911*) or telephone polls (*900-services*).

The various *Call Forwarding* features allow calls intended for a particular line to be routed to another specified directory number, at the terminating party's request. The called party with *Call Forwarding* features violates not only the calling party's one-number-one-line model (since now dialing different numbers may reach the same line), but also the calling party's one-number-one-entity assumption (since the caller may reach a total stranger).

Other features such as those for *Multiple Directory Number Line with Distinctive Ringing (MDNL-DR)* permit more than one number to be assigned to a line, each with its own distinctive ring. As a result, the one-line-one-entity bond at the terminating end is destroyed, even though the one-number-one-line view at the originating end is still preserved.

The features mentioned above have one thing in common: all of them introduce "aliases", or additional numbers (names) for accessing the same line or the same entity, into the telephone numbering system. The alias may be created explicitly by features like *MDNL-DR*, or implicitly by the various routing features like *800-services* or *Call Forwarding*. Consequently, the intent of a feature F that determines the logic of establishing a call based upon either the originating number or the terminating number can be easily circumvented, if F coexists with any of above features that can provide an alias for the number used as F's parameter. A simple case is between *POTS* and *Call Forwarding*, where a POTS call from **A** to **B** may be forwarded to **C**, who perhaps shouldn't know that **A** calls **B**. Examples 11 and 12 are also typical interactions of this type.

3.1.2 Assumptions about Data Availability

A large amount of data is maintained by the network. Some features may assume that certain kinds of data are widely available; other features, such as *Calling Number Delivery Blocking*, may make the same data private. Network capabilities may also limit the accessibility of some corporate data. Features cannot work properly unless they are able to obtain the data needed by their functionalities. For example, even with the *Calling Number Delivery* feature, the device cannot show a caller's number that is not available (this would happen if the call originated from an area lacking the equipment to transmit it). Whether the device should show an unlisted number is still a subject of debate (cf. Example 16).

3.1.3 Assumptions about Administrative Domain

An administrative domain is a telephone network administered by a single organization. A feature that is defined in one administrative domain may not be usable from another administrative domain. One example of this can frustrate American tourists in Europe. Numbers with an 800 prefix are often the only ones widely known for the services of certain American companies. Unfortunately, these numbers cannot in general be reached from the domain of a European telephone company. Reasons for this are that it may be difficult for European telephone companies to bill those American companies for 800 calls, or that American companies may not want to subsidize expensive overseas calls.

Billing is one of several OA&M systems affected by the divestiture of Bell system.

A common perception about billing is that a caller will not be billed for a call unless the call is completed. When there are multiple administrative domains, the definition of a "complete call" can be confusing. Regional companies, interexchange carriers, even service providers of hotels and resorts, may all participate in connecting a simple call. Features that are billed on a per usage basis now need to be specific about the exact basis for a charge. Examples 19 and 20 show two such situations, where the involvement of several administrative domains interferes with the desired execution of billing features.

3.1.4 Assumptions about Call Control

Call control refers to the ability of a user to manipulate (e.g., accept, refuse, terminate, or change the status of) a call. Features such as *911* establish a master-slave relationship—once the call is made, only the *911* operator can terminate it. Others such as *Call Waiting* and *Three-Way Calling* allow a subscriber to put other parties on hold. Another type of control, introduced by several advanced features, is to screen (e.g., *Terminating Call Screening*) or selectively forward (e.g., *Selective Call Forwarding*) some incoming calls. Interactions arise when the call control of one feature prevents other features from exercising their control, as seen in Examples 1, 3 and 4. In Example 4, the *Automatic ReCall* feature alters the role of a user from a callee to a caller, which prevents the user from activating terminating features such as *Terminating Call Screening*.

3.1.5 Assumptions about Signaling Protocol

There used to be a simple protocol that provided the status of the called party when a call attempt is in progress to the calling party: a busy signal means the terminating line is in use; a ring-back means the line is available. Many features now can make a line appear to be available (e.g., *Call Waiting*) or in use (e.g., *Make Busy*). Interactions will occur when one feature needs to know the status of the other party, but cannot tell exactly what the status is from the signal received.[9] Example 13 represents one of them.

New signals introduced by advanced features can also create interactions. A well known case where a feature can interfere with its environment is the use of a modem in a line with the *Call Waiting* feature: an unanticipated call-waiting tone signal can disrupt the data connection. If we view a modem as a third-party service, then this is an example of a feature interaction between a switch-based feature and a third-party service.

3.2 Limitations on Network Support

Network components have their own limited capabilities in communicating with other network components or processing calls. As a result, two seemingly independent features may conflict over the reception of the same signal or the usage of the same functionality.

---

[9]We note that this type of interaction is caused by insufficient availability of *state* data and is in that sense related to the interactions described in Section 3.1.2.

3.2.1 Limited CPE Signaling Capabilities

The set of signals that may be sent from most current CPE's to a switch is limited to *, #, the ten digits 0-9, flashhook, and disconnect. However, with the creation of many new services, the customer needs to send signals for a variety of purposes to the switch. Since there are so few signals available, the same signal is used to mean different things in different contexts. Interactions arise when the interpretation of a signal is ambiguous, as the signal has different meanings in the context of different concurrently active features (see Example 2).

3.2.2 Limited Functionalities for Communications among Network Components

The introduction of common-channel signaling enables various network components to exchange information needed for distributed application processing (e.g., inter-process query and response, or user-to-user data) [11]. Because of the format of the signaling messages and the restricted usage of channel bandwidth for these messages in some network components, features are forced to compete for bandwidth and thus interact. One such instance arises in the realization of AIN features [4]: a switching element must recognize that it needs to send queries to an SCP or an adjunct to trigger AIN features for processing calls. However, a switching element that imposes a "N-query-per-call" constraint on originating queries to an SCP or an adjunct causes interference between different features. This is illustrated in Example 5.

A related problem is the limited information that a SCP can send to a switching element via a TCAP message. A three-digit call type code is used by the SCP to signal the usage of some AIN feature; the switching element then uses the call type code to bill the usage of that feature. When multiple services are invoked by one SCP query, it is not clear how they will be billed (cf. Example 21).

In general, these interactions bring up issues in the protocol and interface design. In the case of launching queries, the protocol for transporting queries to SCPs supports a limited number of queries per call, but the protocol defined by a collection of features requires more; in the case of per usage billing, the interface is not rich enough to carry sufficient information.

3.3 Intrinsic Problems in Distributed Systems

Telecommunications systems are huge, real-time, reactive, distributed systems. Many difficulties in dealing with large distributed systems are also present in managing feature interactions. One obvious case is the problem of *resource contention*. Additionally, the distribution of feature support in the network and the customization of features by each individual can create interactions that require coordination. Some sources of interactions induced by distribution are discussed here.

3.3.1 Resource Contention

The activation of one feature can make other features unavailable. As we mentioned following the discussion of Example 2 on *Call Waiting* and *Three-Way Calling*, the exclusive use of a three-way bridge by one feature automatically disables the other feature.

3.3.2 Personalized Instantiation

Many features allow (or require) each subscriber of the features to assign values to some feature parameters before using the features. A subscriber of *Call Forwarding*, for example, needs to provide the directory number to which incoming calls will be forwarded. For various *Call Screening* features, subscribers have to supply the list of numbers to be screened.

The assignment of values to feature parameters defines the subset of the numbering plan in which the feature is interested. In many cases a particular set of assignments can make two otherwise independent features collide with each other. Examples 9 and 17 illustrate this type of interactions.

3.3.3 Timing and Race Conditions

Timing (i.e., at what time a particular event occurs, or for how long an event lasts) is always critical in distributed systems. The problem is even worse in telephone services, as the phone network also needs to deal with external human behaviors (e.g., dialing numbers, pushing buttons). As a simple example, how long the switchhook is pushed can determine whether it signifies a flashhook or a disconnection. Also, how soon a customer dials the second digit after a dialed "0" could change an intended *0+-service* call into an *Operator-service* call. Since the control context may switch from one feature to another during the call processing, the interpretation of certain signals may depend upon which control thread is active when the signals arrive at the network component. One such interaction is shown in Example 7.

Communication delays also contribute to the interactions in this category. Since signals cannot travel through the network instantaneously, users as well as network components may not have a consistent view of the system status. Together with the timing of events, communication delays can create race conditions that range from nondeterministic event sequences to infinite loops. Examples 2 and 18 illustrate race conditions.

3.3.4 Distributed Support of Features

In the past, one could assume that all features accessible by a single CPE were supported by (or at least known to) the same network component. Therefore, the network component would be aware of all the feature logic, and could coordinate the execution of various features. Unfortunately, this is not necessarily true for the present network architecture, as different features may be supported by different network components even when the features are used by a single user. Consequently, activating one feature in one component may prevent processing of features in another component. Example 6 is one such case.

The AIN architecture further distributes the support of features. In addition to the switch-based (or non-AIN) features that reside in switching elements, there are now AIN-based features whose service logic programs can be in an SCP, an adjunct, or any other intelligent network component. The switching element not only has to know when to launch a query (and where to send the query) to trigger some AIN-based features, but also needs to anticipate the outcome of such processing to be sure that non-AIN features will not be bypassed. An interaction of this type is shown in Example 8.

		Nature of Interactions				
Causes of Interactions		SUSC	SUMC	MUSC	MUMC	CUSY
Violation of Assumptions	Naming		8	10	11,12	
	Data Avail.				16	
	Admins. Domain					19,20
	Call Control	1,3,4			14,15	
	Signaling Protocol				13	
Limitations on Network Support	CPE Signaling	2	7			
	Func. of Comm.	5				21
Problems in Distributed Systems	Res. Contention	2				
	Instantiation	4		9	17	
	Timing & Race	2	7		18	
	Feature Support		6,8			
	Non-atomic Op.					22

Table 1: A summary of the categorization of the examples presented in this paper

3.3.5 Non-atomic Operations

One consequence of a distributed service logic environment is that provisioning could be a non-atomic (i.e., divisible) operation. For example, AIN-based *900 Call Screening* gives flexibility beyond the switch-based counterpart, offering additional features such as screening only during certain time periods and screening with PIN override. In the provisioning of AIN-based 900 service, updates must normally be made to two network components, the switching element and the SCP (or adjunct). How these updates are carried out in sequence may affect a customer's ability to place a simple POTS call (see Example 22 for details).

4 Discussion

Table 1 summarizes what we have presented in the previous sections. Some examples appear in more than one row, because they may have multiple causes, or they may occur only in the presence of more than one cause. Even though this table is not yet complete, it may be difficult for some empty slots to generate an entry. For example, violating naming assumptions is unlikely to result in SUSC interactions, whereas distributed support of features may cause interactions that only tie in with multiple network components. To provide a benchmark as a common measure for coverage, we welcome contributions from other researchers in the field to complete this table, and appreciate suggestions for improving the categorization as well.

4.1 Approaches to Managing Feature Interactions

Currently, there seem to exist three different groups of approaches to managing feature interactions. The first group of approaches is directed at developing infrastructures for the deployment of features in telecommunications systems. The second group supports the design of features (before they are deployed). The third group supports the resolution of interactions as they occur after deployment of the features. Each of these groups of approaches has to deal in some way with the avoidance, detection, and resolution of feature interactions, although typically one group focuses more on one of these aspects than on the others. The following sections describe these groups in more detail.

4.1.1 Infrastructure for Deployment

There are several possible ways to deal with some of the causes of feature interactions. By dealing with the causes of certain feature interactions, the corresponding interactions can be avoided. For example, a new naming scheme with a precise reference either to generic names or to aliases in each feature could avoid violations of naming assumptions (cf. Section 3.1.1); a richer set of functional signals could help resolving some ambiguities caused by limited CPE signaling capabilities; both a standardized application programming interface and a carefully designed interface protocol are useful means to address communication needs and interoperability issues among network components; and a good distributed system platform could manage problems due to non-atomic operations and the distributed nature of feature support in an advanced telecommunications network. These problems are being addressed in three Bellcore projects: AIN [4], INA [12], and the Touring Machine^{TM} project [3].

4.2 Design Support

A second group of approaches typically concentrates on the detection of feature interactions during the design phase. The difficulty of detecting interactions varies. Some interactions can be detected easily by checking the assumptions of a feature against characteristics of the environment in which the feature will be deployed. The detection of interactions caused by limited CPE signaling capabilities is also easy, simply by identifying the set of features that could respond to the same signal at the same time. Others, such as the interactions caused by timing and race conditions, may require sophisticated formal techniques for specification and reasoning [6, 10]. Techniques for protocol engineering, including design, specification, verification, validation, and testing, will be very useful in detecting interactions in this area.

4.3 Run-time Resolution

Some causes of feature interactions may be difficult to deal with at the feature design phase. Because of diverse preferences of customers, no single policy governing the availability of data could be satisfactory, nor could a set of rigid precedence relations exist for resolving conflicting call controls. Instantiations of feature parameters may be so unpredictable that dealing with all possible cases before the deployment of features could be infeasible. Differences in objectives also create barriers to any kind of agreement among administrative domains. A run-time resolution scheme based upon techniques developed in the area of distributed artificial intelligence is a possible approach [8].

5 Concluding Remarks

The feature interaction problem arises in many phases of service creation and deployment. However, while talking to individuals with different experiences and involvements in the process of service creation, we were surprised to learn that the problem conveys so many images: viewed by electrical engineers with a signal processing background, feature interactions are as simple as "signal ambiguities", and can be resolved easily using "functional signaling"; viewed by requirements writers, feature interactions

---

^{TM} Touring Machine is a trademark of Bellcore.

are inevitable consequences of ambiguous or incomplete specifications/requirements of telecommunications services, and tools to aid in the creation of concise, accurate requirements are crucial; viewed by software developers, the problem is similar to software engineering issues such as *extensibility*—the needs of new services have not been anticipated by the designers of the existing system; viewed by distributed system researchers, how to manage the issues of naming, control strategies, and race conditions is critical; and for artificial intelligence researchers, feature interactions occur as conflicting customer needs and/or policy objectives that can in many cases be resolved by negotiation.

While these viewpoints all represent fundamental concerns in the creation of interworking features, our categorizations and examples provide a tangible "check list" of specific issues to be dealt with.

A successful advanced service platform depends upon identifying adequate ways of managing feature interactions. This in turn requires an innovative resolution framework and a powerful interaction detection mechanism. A comprehensive understanding of the causes of adverse feature interactions is both fundamental and essential.

Acknowledgments

The authors thank Frank S. Dworak and Gary Herman for many insightful discussions. We are also grateful to Ron F. Baruzzi, Ralph B. Blumenthal, Ritu Chadha, James E. Katz, Martin Krupp, Shoshi Loeb, Ashok Ranade, and Brian S. Whittle for useful comments.

References

[1] *LATA Switching Systems Generic Requirements (LSSGR)*, Bellcore TR-TSY-000064, FSD 00-00-0100, July 1989.

[2] *Proceedings of the 3rd Telecommunications Information Networking Architecture Workshop (TINA-92)*. Narita, Japan, January 1992.

[3] M. Arango, P. Bates, G. Gopal, N. Griffeth, G. Herman, T. Hickey, W. Leland, V. Mak, L. Ruston, M. Segal, M. Vecchi, A. Weinrib, and S.-Y. Wuu. Touring Machine: a software infrastructure to support multimedia communications. In *MULTIMEDIA '92, 4th IEEE COMSOC International Workshop on Multimedia Communications*, Monterey, CA, April 1-4 1992.

[4] Advanced Intelligent Network Release 1 Proposal. Special Report SR-NPL-001509, Issue 1, Bellcore, November 1989.

[5] Generic Framework Criteria for Wireless Access Communications Systems (WACS). Framework Technical Advisory FA-NWT-001013, Issue 3, Bellcore, March 1992.

[6] E.J. Cameron and Y.-J. Lin. A Real-Time Transition Model for Analyzing Behavioral Compatibility of Telecommunications Services. In *Proceedings of the ACM SIGSOFT '91 Conference on Software for Critical Systems*, pages 101–111, December 1991. New Orleans, Louisiana.

[7] F.S. Dworak, T.F. Bowen, C.H. Chow, G.E. Herman, N. Griffeth, and Y.-J. Lin. Feature Interaction Problem in Telecommunication Systems. In *Proceedings of the Seventh International Conference on Software Engineering for Telecommunication Switching Systems*, pages 59–62, July 1989. Bournemouth, United Kingdom.

[8] N.D. Griffeth and H. Velthuijsen. The Negotiating Agent Model for Rapid Feature Development. In *Proceedings of the Eighth International Conference on Software Engineering for Telecommunications Systems and Services*, March 1992. Florence, Italy.

[9] A. Lee and D. Carrington. Formalizing Extensions and Modifications to Telecommunication Software. In *Australian Conference on Telecommunications Software (ACTS)*, pages 205–210, 1991.

[10] Y.-J. Lin. Analyzing Service Specifications Based upon the Logic Programming Paradigm. In *Proceedings of the IEEE GLOBECOM '90*, pages 651–655, December 1990. San Diego, California.

[11] A.R. Modarressi and R.A. Skoog. Signaling System No. 7: A Tutorial. *IEEE Communications*, pages 19–35, July 1990.

[12] H. Rubin. Tutorial on Information Networking Architecture. In *Proceedings of the IEEE GLOBECOM '91*, 1991.

Using an Architecture to Help Beat Feature Interaction

Rob van der Linden
ANSA Project, Architecture Projects Management Limited,
Poseidon House, Castle Park, Cambridge CB3 0RD, UK,
rvdl@ansa.co.uk

Abstract. Results of work on open distributed processing systems appear to be applicable to the solution of certain feature interaction problems in telecommunications systems. The principles of *separation* and *substitutability* are instrumental in application independent structuring of systems. The provision of extensive information about systems and features is also shown to be of great importance. An architectural framework for organising this information is proposed. Some mainly organisational issues, which stand in the way of making this information widely available, are identified.

1 Introduction

Telecommunications service providers are no longer able to limit their product portfolios to a few simple standardised interconnection services, such as POTS and data transmission services. Technological advances are driving the cost of providing high capacity interconnection services down. Increased competition is reducing the scope for substantial profit margins; without the introduction of new services, the market for simple interconnection services is becoming saturated.

To maintain revenue, telecommunications service providers will increasingly have to provide many complex specialised value added services to small groups of individuals. In the emerging fiercely competitive market, customers demand:

- the latest personalised services, e.g. video on demand;
- service provision at competitive rates, e.g. similar to video cassette rental;
- high quality services, e.g. no connection drop-out during movie watching;
- control over service configuration, e.g. no X-rated movies for children;
- some or full integration amongst services offered by different suppliers, e.g. access to several video libraries, using the same controller over different networks.

This demand implies an increase in the number of services, a need for rapid deployment, and with smaller user communities per service, provision needs to be low cost.

To fulfill customer demands an increasing proportion of value added services is provided through software. This adds significantly to the flexibility of the switching and transmission components but increases the complexity of the resulting systems. Software design, installation, testing, versioning, configuration, and maintenance in a widespread heterogeneous environment is becoming increasingly difficult. With each new service and/or feature, systems become more complex. Complex systems are connected together in places where no such interconnections were foreseen. Nobody understands the whole of the system. Nobody knows for sure that additions and interconnections will have precisely the desired effect, and no other.

The study area of feature interaction in telecommunications systems has grown out of this situation [1]. A major challenge appears to have been placed at the door of all those who are involved with the software design and production process. This challenge is not new: much has been published about the design process (from waterfall models to the idea of growing software), many methodologies have been proposed (from structured analysis and design to object orientation), and many approaches have been suggested (from drawing pictures to formal methods with mathematical foundations). Tools have been provided for most of the above. None of these have turned out to be the Silver Bullet they promised to be, perhaps because most approaches tackle the accidental tasks which arise from software production and not the essence of fashioning the appropriate conceptual constructs [2], [3].

In the area of open distributed processing much has been done to redress this balance [4], [5]. In particular, two ways of reducing the complexity of distributed computing systems and their design have been pursued:

- Place constraints on the structure of systems, their software components and the ways in which these may interact, such that software components can be separately analysed and replaced.

- Provide a framework for describing systems, their software components and the interactions between them in a way which is independent of the process by which such systems are designed and built.

The principles of *separation* and *substitutability* are described in the next section and can be shown to be powerful weapons in the reduction of feature interactions.

A framework for organising information about systems and their components is then described. Unlike most other approaches it does not enforce the traditional division of labour and the definition of roles of those involved in the business process of software development and production.

Cameron et al. [6] identify three main causes of feature interactions: violations of feature assumptions, limitations of network support, and intrinsic problems in distributed systems. The ideas expressed in this paper mainly tackle the latter, and recognise that providing extensive information about systems and features contributes to the former two.

2 Structuring distributed systems

Bowen et al. [1] describe feature interaction issues in relation to each of the phases of the development process. In the design environment, they cite the problem of subverting the original system structure by introducing new features. In the implementation environment they refer to the problem of reuse, where seemingly minor changes are made to a software component to meet a new need. As a result, feature interactions may be introduced.

These are real problems in systems where system structure primarily depends on the application in question, and where designers have the freedom to structure the required functionality in any way they see fit. The basis for taking decisions on a particular system structure will often also be based on assumptions about the specific technologies which are (or must be) used for the implementation of the system and on the personal views of the designers. Such assumptions and views are usually undocumented or lost beyond the design phase. Most importantly, they are inaccessible to those who have to add new features at a later stage.

The open distributed processing community has derived two important principles to deal with this problem: *separation* and *substitutability*.

2.1 Separation

When two entities (features) are completely physically separated then there is no way in which the two can interact in either a positive or negative manner. This characteristic is exploited when constructing secure systems. Of course it is desirable to have entities interact in a controlled manner; controlled such that only desirable interactions take place. To achieve this it is necessary to bring the entities together just enough for the desirable interactions to be possible. By minimising the extent to which components are brought together it is possible to minimise the scope for introducing undesirable feature interactions.

In open distributed computing, software components are encapsulated and allowed to interact only through well defined interfaces. The encapsulated components include their private memory, threads and schedulers for concurrent operations. They each carry most of their communications protocols with them as well. They thus rely only on minimal shared communications and processing resources. The principle of minimal sharing translates into minimal undesired feature interactions (or side effects in software speak).

The unexpected benefit from this approach is improved portability. As software components depend less on the now simple infrastructure, the task of making a component operate on an infrastructure based on different technology is made a lot simpler.

There is a hidden cost in that every software component now has to include quite a bit more functionality. A reduction can be achieved by carefully engineering the components. As a principle, only those mechanisms which need to be included should be included. It makes no sense to have full multi-threading support available in a component which allows no concurrent access, or dependability mechanisms in non-critical components. Furthermore, components which are always located in one place (e.g. a process) do not need support for remote communications for interactions in the local context. Software tools can be employed to include just those mechanisms which are needed. The required mechanisms are indirectly specified by the designer who states the required properties of each component in the system. Thus human error is removed as designers no longer have to deal with "system functions" in detail. This further contributes to the reduction in feature interactions.

2.2 Substitutability

The interactions between software components in open distributed computing have also been stylised. The separation principle dictates that components can only interact through previously defined interfaces. This affords a great deal of control over the (legal) interactions. For example, in ANSA there are two kinds of typed interfaces:

- Operational interfaces, which contain a set of *named operations*, each with a set of zero or more typed *parameters*. Each operation can result in one of several *outcomes* or *named terminations*, each with zero or more typed *results*. These are used for the usual procedure call style interactions.

- Stream interfaces, in which communications is structured as a set of linked directed information flows. These are used for computationally unstructured communications, e.g. video and voice streams in multi-media applications.

The operational interface type system is reminiscent of type systems in programming languages. Type checking, the process of checking the safety of interactions between software components is based on the rule of no surprises. The entity which receives an operation invocation (the server) should never be surprised by an unexpected operation, or unexpected parameters. The entity which invokes the operation (the client) should never be surprised by an unexpected outcome, or unexpected results. This rule allows a server to expect more operation invocations than a client can produce and a client to expect more outcomes than a server can produce.

The no surprises rule should be applied whenever and wherever two software components are to interact. Type checking takes place as early as possible, so as to detect potential errors as early as possible. In static systems, type checking is part of the compile and link process. Often, type checking information is lost from that point onwards. In open distributed processing systems, the type information which describes the interface is preserved and stored, for instance in an interface repository [7]. In dynamic systems, software components come and go, as do their interface descriptions. Even in that situation, type checking can be performed before interactions actually take place.

Once a software component needs to be replaced or extended, it can be substituted by a component which is *type compatible*. If the type information is available then there is less chance of mismatches and subsequent feature interactions after the change has taken place. Note that keeping just type information is not sufficient, as will be demonstrated in a next section.

Bowen et al. [1] display concern about the modification of existing functions to meet the requirements of a new feature. This "re-use" can cause changes which "ripple" through to cause catastrophic failures of other features. The danger of this situation occurring, given a system in which software components are separated and in which interactions are stylised and strongly typed is much reduced. It should in any case be easier to analyse and detect such ripple effects if the system is well structured.

2.3 Relationship with CCITT IN

The relationship between open distributed processing systems and Intelligent Networks may not at first appear to be obvious, beyond saying that IN includes a good deal of processing and is indeed a distributed system. Furthermore the draft open distributed processing standards [5] have been developed in close cooperation with ITU-T (formerly CCITT).

On a more technical note, the intention is that stream interfaces are combined with one or more operational interfaces in a single component. The operational interface then offers a set of operations by which the stream interface can be controlled, particularly in terms of the way in which that interface is connected to other stream interfaces. Such an object displays similar functionality to the combination of a Service Switching Function (SSF) and a Call Control Function (CCF) in IN-CS1.

Thus it becomes possible to build software applications, similar to Service Logic Programs (SLP) in IN-CS1, which control, switch and otherwise manage streams of computationally unstructured information. The open distributed processing view however does not insist on the presence of a Service Control Function (SCF), which behaves as a shared resource for a set of SLPs since such a shared resource could have feature interaction implications.

The separation principle in open distributed processing systems, would encourage the separation of data in the Service Data Function (SDF) such that data which changes frequently, but has no concurrent access (e.g. billing data) would be separated from data which is stable but read frequently and often concurrently (e.g. for number translation). This would allow both billing and number translation data components to be without expensive locking mechanisms for instance.

The encapsulation principle would encourage designers to place data with the programs which access that data. The strict separation between SCF/SLP and SDF would therefore be too rigid.

A more general point is that CCITT IN proposals are for the first time separating the concerns of the network providers from those of the value added service providers [8]. Open distributed processing principles will therefore be much simpler to apply in IN than in many existing more hybrid telecommunication switching systems.

2.4 Guidelines for structuring systems

A system or feature may be thought of as consisting of:

- a set of (software) components;

- a set of interactions between these components;

- information about the composition of each of the (software) components.

This can thus be highly recursive and, depending on the number of components, the information about them can quickly overwhelm.

The properties of software components and the nature of their composition must therefore be constrained to ensure that systems built out of allowable components can be analysed without danger of an exponential explosion of combinations. Restricting the set of software components to a very small set however has a negative effect:

- software components may be too large for most applications;

- software components may be used in inappropriate ways, perhaps because the available ones do not display sufficient functionality;

- a particular software component will be used for more than one function, leading to lack of separation of function and resources and a danger of feature interaction.

Furthermore, software components should not be specified in a predefined structure, similar to a system design. This will overly restrict their application. The result may be that components will be used in ways or places where they were not intended to be used. For example, genericly describing a system in terms of several layers of functionality often leads to system implementations in which each of those layers is actually represented, even if their presence serves no purpose. This of course leads to large and often slow software and is also a common cause of feature interactions.

What is needed is a software architecture. Such an architecture [4], specifies:

- a set of software components, which form the basic building blocks and tools of the architecture;

- a set of rules, which constrain the way in which the components can be combined in designs which conform to the architecture;

- a set of recipes, which provide advice on how to combine basic components, using the tools to obtain subsystems with certain properties;

- a set of guidelines, which help designers make design decisions if their own preferences do not offer sufficient guidance for such decisions.

Using the components, rules, recipes and guidelines (i.e. the architecture) helps ensure that the differences between (sub)systems which each conform to the architecture remain such that services can be ported from one technology to another and that their interworking can be achieved in a controlled manner. In particular this means that the effect of interactions between components can quickly be analysed.

This has not helped much in solving the problem of what components should actually be chosen by designers of features in telecommunications systems. To answer this question requires much more knowledge of actual features and telecommunications applications in question. However, the general guidelines above do guide the choice of a particular set of software components and help ensure that the task of reasoning about relationships between them is simplified, because any interactions are stylised in an architectural framework.

An early example of the use of distributed computer systems ideas in telecommunications systems was published by Mierop et al [9]. Research in this area is continuing [10].

3 Information about features

To reduce feature interactions it is necessary to provide extensive information about systems and features. Two kinds of information can be distinguished. First there is information about a feature being activated in a running system. For instance, when A redirects calls to B, then B needs to keep information about this in case of further redirection from B to C say. B may refuse this or refer such redirection back to A for instance (all in an attempt to avoid a circular redirection path). In this section a second class of information receives attention: the information which describes the software components themselves. This information is typically generated during the design and development process. It is needed for modification of existing features and for rapid deployment of new ones.

Traditionally this information is structured to be homomorphic with the structure of the design and development process. That is, requirements information, specifications, design details, implementation and maintenance information are all kept separately and quickly become inconsistent.

Many companies are organised in ways which encourage this separation between specifiers, designers, implementors, testers and maintenance personnel. This separation does nothing to reduce the scope for introducing feature interactions:

- Specifications, designs and implementations are handed from one department to another, sometimes even from one company to another. Such handovers are often manual and paper based. The consequent information loss causes a certain ignorance about aspects of the system under consideration. Insufficient knowledge about the system in turn allows specification, design or implementation decisions which result in feature interactions.

- Designers and implementors are often forced to deviate from a specification when practical design or implementation issues are under consideration. They are unable to record such deviations with the original specification. Undocumented deviations cause major problems during testing and are a frequent cause of feature interactions.

When specification of new features takes place in the context of an existing system, then information about this system must be complete, unambiguous, and easy to understand. The design of a feature takes place in the context of a specification for the new feature and knowledge about the design of the system in which the new feature is to be added. Similarly, implementation takes place in the context of a specification, a design and the actual system in which the feature is to operate. At each stage it is necessary to have access to as much information as possible. At each stage it may be necessary to make minor modifications to aspects of previous stages. In that sense there is no such thing as an exact order to design, development and implementation stages.

Modern tools help bring some consistency to specification, design and implementation information, which is stored in their repositories. These tools do not support very large system designs, or very large geographically distributed teams.

Telecommunications systems are inherently complex. To cope with these complexities, it is necessary to separate the concerns of those involved and deal with these within an architectural framework which provides overall integrity. It follows that the information about systems and features should be separated in a similar way.

The process of designing and building systems within the architecture is constrained only by the nature of the components and the composition rules. This is because architecture applies to components and their configurations and is principally structural. The way in which a conforming structure is designed and built is of secondary interest. An architecture therefore should not prescribe any particular design process or methodology. Consequently, it makes no sense to classify the information about features and systems according to the stages in a design process. As stated previously, separation between stages in the design process is already a cause of possible feature interactions.

In addition, it is desirable to make all those involved in the design process acutely aware of the reasons for their activities: to provide some end-user functionality.

The distributed processing community (aka. ODP) has derived a framework which it claims can capture all issues associated with a particular design [5]. It aims at completeness and consistency of designs and implementations. The description of software components and their interactions (see previous section) can thus be divided into statements about:

- the purpose of the software component;
- the meaning of the interactions in which the component may take part and the intended use to which the component may be put;
- the structure of the component in terms of distributable functions and interfaces;
- the quality of the component and a statement of how this may be achieved in the distributed environment;
- the technology which is applied in realising the software component.

The statement of purpose of components may include details about security, privacy and management policies to which the component adheres, and any accounting, billing, auditing and monitoring arrangements which may apply. All these are described in the context of obligations and liabilities of stakeholders (i.e. the "owner", "provider" and "user" of the component). Descriptions may be couched in natural or semi-natural language. This is easy to understand and generate, but not easy to make consistent. Requirements or process modeling languages may also be used.

The meaning of interactions must be separately specified because parameters and results in each interaction should be interpreted as references to other (software) components. Which components can be named, the context in which naming proceeds and any naming conventions must be specified to help refer to the reason for the component's existence and the meaning of its interactions. There exists very little linguistic support for the description of these aspects of a component.

Each software component must be described in terms of its own structure and the structure of its interactions. This includes any constraints on ordering of operation invocations, any constraints on parallel execution of algorithms, and ways in which results are computed. Most specification and programming languages are suitable for this kind of component description. Specialised interface definition languages (IDL) have been developed to describe interfaces in a programming language independent manner. Many formal languages exist and are used to help bring rigour to the software production process.

Software components may be engineered to offer different quality of service. Their dependability or performance characteristics may vary depending often on the quantity and quality of the resources used to construct them. The quality of service of each component must be explicitly stated. This may be done in the context of (with reference to) specific technologies (e.g. replication and hot stand-by). This part of a component description will guide the trade-offs which each designer must face. For instance, making a component more available by replication, will require mechanisms to deal with possible inconsistencies amongst replicas. The same applies for interactions with other components. Examples of specific quality of service parameters here are bandwidth, error rate and jitter. There exists little specialised linguistic support for these descriptions.

Finally, each software component, if it is to be realised, must make use of technologies, such as networks, switches, operating systems, signaling protocols, messaging systems and so on. Component interactions can be specified by (reference to) protocol definitions. This part of a components description is sometimes explicit through references to libraries, and compiler, linker and/or loader options. Today, a lot of this information is lost once the component has been constructed and loaded in a system; some of it can be found back by reference to make files etc.

4 Available descriptions and accessible software

Having derived as complete a description of a software component as possible, it is necessary to ensure that this description becomes available to all those who need to have access to it. In general this means everyone who is in some way legally interested in the component and its interactions. Maintenance personnel will need to know about the components they are maintaining. Designers of new features will require access to descriptions of the components already in the system. They will also be extending descriptions of components which make up a new feature.

At times they will want to query the set of descriptions to see if a component with desired characteristics has already been developed. This raises the question of what kinds of query should be supported. Using formal languages is probably not a good idea [11]: writing the query would take as long as writing the specification, syntactically different descriptions may be semantically the same, and there is no such thing as a close match between two formal specifications.

Designers will also want to get access to completed software components, or to organisations which are proposing to build such components, once they have identified the required characteristics. In open distributed computing systems this can be provided in one of two ways: either a reference to a construction service (a factory), or a reference to an actual instance is passed back to the party that invoked the query. A construction service is able to construct a software component with particular characteristics.

To support these and other design, development and maintenance activities, on a large scale, the information service must itself be built as large scale service in a wide area communications network. (The structuring principles discussed earlier in this paper should of course be applied to such a system.) One prototype, under construction in ANSA, combines existing technologies, such as traders [12], interface and implementation repositories [7], and relational database technology together to offer an extendible information service. Such a service needs to span organisational boundaries and be able to store a wide variety of information. In the future this information will also be used to drive application transformation and configuration tools.

Several other obstacles also have to be overcome specially when software components and information about them are passed across organisational boundaries. The problems are partly technical, partly human, and partly a consequence of the way in software is procured. Wood and Sommerville [11] point out that the human problem is a result of the fact that software reuse involves a kind of de-skilling of what are seen as highly skilled developers. The economic problems are associated with the problem of deciding who owns a complex object which has been built by X for Y, using components developed by Z. The technical problems are associated with the difficulty of developing reusable components, the difficulty of a priori deciding whether a component is actually reusable, and the absence of any effective catalogues, which can be searched effectively. Current work at ANSA is addressing some of these issues.

There are many other differences on either side of the organisational boundary:

- No common purpose; different organisational entities often lack a common purpose, specially if boundaries lie between dissimilar industries.

- No trust; there is an inherent lack of trust between different organisational entities. This is most vividly reflected in the uncertainties about the quality of a software component which is obtained from another domain.

- No common economic framework; where components are provided by one organisation and used in another there may be a need for payment. There is no clearly defined market for software components or information about them, neither is there a clear idea of how to charge for either use or ownership.

- No common legal framework; software components may be of inferior quality, or the information about a component may be inaccurate. It is unclear what if any legal framework applies in these cases.

- No common meaning and representation; there is little agreement about the languages which should be used for the description of software components. Misunderstandings can arise despite the presence of comprehensive descriptions (some aspects can only be described in natural language for instance; formal languages need a common context for agreed interpretation).

- No common quality standards; the agreed measures for software component quality have been derived from those suitable for hardware components (MTBF, MTTF etc.). Software quality of service is often better expressed in application dependent terms, such as mission time (the time during which a component is to stay operational with a precisely defined probability of failure for instance) [13].

- No support for distribution; this is a problem which requires the application of distributed processing technologies.

The above organisational issues should be solved within the business process of producing new features for telecommunications systems. They are part of a business process reengineering activity [14] in which several telecommunications companies may wish to become engaged.

5 Summary

In response to increased competition and more varied customer demands, telecommunications service providers have to increase the flexibility of the services and features they provide. They also have to become more efficient at rapidly providing new services and features and alterations to existing ones.

At least two principles, borrowed from work on software architectures for open distributed processing systems are relevant in reducing the chance of feature interaction. The principle of *separation* allows systems to be structured in ways which reduce the chance of feature interactions which are introduced due to unexpected resource sharing. The principle of *substitutability* allows explicit checks to be made whenever two software components are to mutually cooperate through a set of required interactions, thereby at least reducing and perhaps eliminating unexpected mismatches.

To further reduce the scope for introducing feature interactions, it is important to make descriptions of software components widely available and make the components themselves widely accessible. Much specification and design information is available but either lost during the development process or separated from the run-time system in some other way. The consequences of this loss of information are duplication of effort, long lead times and in telecommunications systems, unwanted feature interactions.

It is important to understand how the necessary information may be derived, organised, and how it can be made available as widely as possible. If the latter can be achieved, then specifiers, designers, implementors, testers and maintenance personnel will take best advantage of the information provided. They will become able to take better informed decisions, document these and thus reduce the scope for feature interactions.

Acknowledgements

The author wishes to thank E. Jane Cameron (Bellcore), Andrew Herbert (APM) and Neil Mason (GPT) for their insightful comments.

References

[1] T.F. Bowen, F.S. Dworak, C.H. Chow, N.D. Griffeth, G.E. Herman and S.-J. Lin, The Feature Interaction problem In Telecommunication systems, *Proc. 7th International Conference on Software Engineering for Telecommunications Switching Systems*, pp 59-62, Bournemouth, UK, July 1989.

[2] F. Brooks, No Silver Bullet, Essence and Accidents of Software Engineering, *IEEE Computer*, Vol 20(4), pp. 10-19, April 1987.

[3] D. Harel, Biting the Silver Bullet, *IEEE Computer*, Vol 25(1), pp.8-20, January 1992.

[4] R.J. van der Linden. An Overview of ANSA, Architecture Report, Architecture Projects Management Limited, Cambridge, UK.

[5] ISO/IEC 10746, ITU-TS Recommendation X.900: Basic reference Model of Open Distributed Processing, Draft International Standard, March 1994

[6] E.J. Cameron, N.D. Griffeth, Y.-J Lin, M.E. Nelson, W.K. Schnure and H. Velthuijsen, A Feature-Interaction Benchmark for IN and Beyond, *IEEE Communications Magazine*, Vol 31(3), pp.64-69, March 1993.

[7] The Common Object Request Broker: Architecture and Specification, Revision 1.1, OMG Document number 91.12.1, OMG, 1991, Revision 1.2, (available from OMG file server), OMG, 1994.

[8] R.L. Bennett and G. Policello, Switching Systems in the 21st Century, *IEEE Communications Magazine*, Vol 31(3), pp.24-36, March 1993.

[9] J. Mierop, S. Tax and R. Janmaat, Service Interaction in an Object-Oriented Environment, *IEEE Communications Magazine*, Vol 31(8), pp.46-51, August 1993.

[10] W.J. Barr, T. Boyd and Y. Inoue, The TINA Initiative, *IEEE Communications Magazine*, Vol 31(3), pp.70-76, March 1993.

[11] M. Wood and I. Sommerville, An information retrieval system for software components, *Software Engineering Journal*, pp.198-207, Sept. 1988.

[12] ISO Open Systems Interconnection, Data Management and Open Distributed Processing, Draft ODP Trading Function, ISO/IEC JTC1/SC21/WG7/N880, November 1993.

[13] D.P. Siewiorek and R.S. Swarz, Reliable Computer Systems, Design and Evaluation, ISBN 1-55558-075-0, Digital Press, 1992.

[14] M. Hammer and J. Champy, Reengineering the Corporation, Harper Business, Harper Collins Publishers, New York, 1993.

Towards Automated Detection of Feature Interactions

Kenneth H. Braithwaite Joanne M. Atlee[1]
Department of Computer Science
University of Waterloo
Waterloo, Ontario N2L 3G1

Abstract. The *feature interaction problem* occurs when the addition of a new feature to a system disrupts the existing services and features. This paper describes a tabular notation for specifying the functional behavior of telephony features. It also describes how four classes of feature interactions can be detected when features are specified in this new notation. The goal of this research is to develop a tool to automatically analyze feature specifications and detect interactions at the specification stage.

1 Introduction

How does one add features to a system without disrupting the services and features already provided? A more difficult but related problem is: how can one ensure that combinations of independently developed services and features behave as expected? These questions, and other variations of the *feature interaction problem*, have plagued the telecommunications industry for several years [7]. More generally, they are problems that affect the development and evolution of all service-oriented software.

We use the following definitions of *service* and *feature* presented in [3]

- A *service* provides stand-alone functionality. For example, Plain Old Telephone Service (POTS) is a service.

- A *feature* provides added functionality to an existing feature or service; a feature cannot operate stand-alone. For example, *Call Waiting* adds functionality to POTS.

- A *feature interaction* occurs when one feature affects the behavior of another.

Sometimes feature interactions are desired, and one feature is explicitly designed to interact with another. In such a case, one wants to determine that the features not only interact, but that the interaction conforms to the specified behaviors of the individual features [14]. In most cases, one simply wants to ensure that (supposedly) non-interacting features cannot interact.

We are investigating how to detect feature interactions during the requirements phase of development. Our goal is to develop a requirements notation that is flexible enough to support the specification of a wide variety of telephony events and properties

---

[1]This research has been supported in part by the Natural Sciences and Engineering Research Council of Canada grant FSP137101, with matching funds from Bell-Northern Research Ltd.

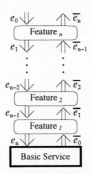

Figure 1: Information flow through layered state-transition machines.

but is rigorous enough to allow automated analysis. Automatic comparison of feature specifications is essential. For example, thousands of features have been implemented for telecommunication systems: the DMS-100[1] switch alone supports over 800 telephony features [10]. The number of possible combinations of telephony features precludes effective comparative analysis by humans. Automated analysis is also important when new features are created by third-party developers, since their designers are not likely to be intimately familiar with the base system and existing features.

At present, we have designed graphical and tabular notations for specifying the functional behavior of telephone services and features, and we have developed algorithms for detecting certain types of interactions among features. Using the taxonomy of feature interactions presented in [4], the classes of interactions we have been able to detect include interactions caused by call control manipulation, information manipulation, and resource contention; violations of some types of feature invariants can also be detected.

In this paper, we describe the current status of our specification notation and the algorithms we have developed for detecting feature interactions. We demonstrate our method by showing how interactions can be detected among the specifications for features *Call Waiting*, *Three-Way Calling*, *Originating Call Screening*, *Call Forwarding*, *Calling Number Delivery*, and *Calling Number Delivery Blocking*. However, we anticipate that our method can be generalized and used to detect feature interactions in other service-oriented systems.

2 Modeling Services and Features

We are primarily interested in studying systems whose behavior can be modeled as layered state-transition machines. In such systems, the zeroth level machine specifies the system's basic services, and higher-level machines specify features that enhance the behavior of lower-level machines (see Figure 1). Information from the environment is input to the top-level machine and propagated down through the layers. At each level n, the n_{th} machine may either pass its input unaltered to the next machine; consume its input and perform some action; or consume its input, perform some action, and produce new information to be passed to the next machine. Examples of such systems include computer networks, operating systems, telephony systems [12], and robotics [2, 6]. The remainder of this paper will concentrate on telephony systems.

---

[1]DMS is a trademark of Northern Telecom.

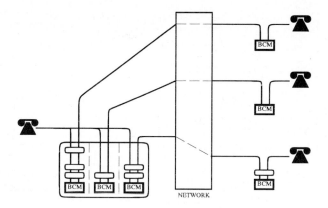

Figure 2: Representation of a call using single-ended call model.

2.1 Layered State-Transition Machine Model

The basic service offered by a telecommunications system is to establish and maintain a communications link between two *agents*, which may be terminals or trunks. Traditionally, services and features have been defined in terms of a call model that encompasses both the originating and terminating agents. The advantage of this architecture has been improved performance, because all of the information a service or feature needs is locally available. The disadvantage is high coupling among the modules which implement services and features. As a result, service and feature modules are tightly interconnected, and it has become increasingly difficult and expensive to maintain and enhance the system.

To rectify the situation, and to emphasize changeability and reusability over performance, telephony systems are being re-designed using a single-ended call model that is based on agents rather than on connections [12]. The representation of a call using the single-ended call model is depicted in Figure 2. The behavior of an agent's call is specified by its *call stack*: a set of layered state-transition machines. The zeroth-level machine of an agent's call stack specifies basic call processing, and the higher-level machines specify activated telephone features (e.g., *Three-Way Calling*, *Hold*, etc.). The behavior of a connection between two agents is the composition of their call stacks. If an agent is involved in more than one two-party call then it has multiple call stacks, one for each two-party call.

State-transition machines depicting the basic call processing service are shown in Figure 3. The state machine on the left specifies the *Originating Basic Call Model (OCM)*; it describes the behavior of a call that is initiated by the agent. The state machine on the right specifies the *Terminating Basic Call Model (TCM)*; it describes the behavior of a call that is being received by the agent. The figures are based on the basic call models defined in AIN Release 1.0.

State-transition machines specifying activated features are layered on top of the agent's basic call model in order of their priority. The machine on the top of the stack has the highest priority because it has the first chance to act on incoming messages and has the last chance to modify outgoing messages.

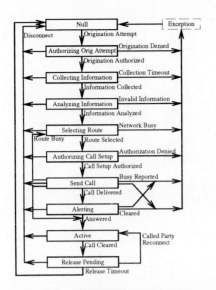
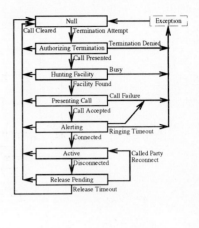

Figure 3: (a) Originating Call Model   (b) Terminating Call Model

2.2 Extended Layered State-Transition Machine Model

Our model is an enhancement of the layered state-transition machine model described above. To improve the readability, modifiability, and reusability of feature specifications, we propose creating separate state machines for the different types of calls the feature can modify. For example, if a feature modifies the behavior of more than one call, then a separate state-transition machine is defined to specify how the feature affects each call. Separate state machines are also created if the call that the feature modifies can either be an Originating Call or a Terminating Call. The types of information passed up and down the call stacks of Originating and Terminating Calls are different, so the input that a feature consumes, outputs, or uses to trigger state transitions depends on which underlying call model the feature is operating on.

Information that is passed up and down the call stack is represented by tokens. Some tokens carry no information other than their name, such as the *flashhook* token. Some tokens have a variable value, such as the *target* token, which represents the directory number of the called agent. Some tokens are compound, having other tokens as components: for example, the *Call Request* contains the set of tokens needed by a terminating agent to establish a connection with the originating agent; *Call Request* includes component token *target*. Two other types of token deserve mention. The first is a *Transition Attempt Notification*. Before a feature or underlying call model can change state it produces a *Transition Attempt Notification* which is passed down the stack from the top. If a feature receives its own *Transition Attempt Notification*, then it has permission from the higher-priority features to transition to the new state. The features discussed in this paper only make use of *Transition Attempt Notifications* from the underlying call models. The other token of interest is a *Feature Signal*. *Feature Signals* are passed only between different sub-machines of a feature whose state-transition machine is distributed across different call stacks.

Consider the telephony feature *Call Waiting*, which allows an agent (subsequently referred to as the *user* of the feature) to accept a call while on the phone with another agent. If a call comes in while the user is on the phone, *Call Waiting* generates a signal

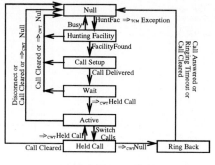

(a) New call

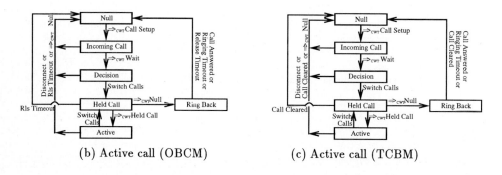

(b) Active call (OBCM)   (c) Active call (TCBM)

Figure 4: Call Waiting feature using split call model.

to notify the user that another call has arrived. At this point, the user has the option of putting the active call on hold and accepting the new incoming call. Figure 4 contains state-transition machine specifications for the *Call Waiting* feature. In Figure 4, the top machine describes how *Call Waiting* affects the behavior of the incoming call. Since the new call must be a Terminating Call, only one state-transition machine is needed. However, the call that existed before *Call Waiting* was activated could either have been initiated or received by the user. Thus, we create two state machines to describe how the feature affects the call that was already in progress when the new call comes in: the machine on the bottom left in Figure 4 describes how *Call Waiting* modifies an active call the user originated, and the machine on the bottom right describes how *Call Waiting* modifies an active call the user received. The only difference between the two state-transition machines for *Call Waiting* on an active call is the token that the call model receives from the communication link indicating that the remote agent has hung up the phone: an Originating Call receives a *Call Cleared* token whereas a Terminating Call receives a *Release Timeout* token.

The graphical representations of state-transition machines shown in Figure 4 provide an intuitive understanding of how the *Call Waiting* feature affects normal call processing. A new call stack based a Terminating Call Model is created whenever a call arrives for an agent. *Call Waiting* is activated when the Terminating Call Model determines that the agent is busy. At this point, the agent has two call stacks: one for the original call and one for the call that just arrived. The *Call Waiting* machine for a new call is placed on the new call's call stack, and the appropriate *Call Waiting* machine for an active call is placed on the original call's call stack.

State	Input	Output	NewState	Assertion
NULL	$\Rightarrow_{CWT}$ CALLSETUP		INCOMINGCALL	
INCOMINGCALL	$\Rightarrow_{CWT}$ WAIT		DECISION	
	$\Rightarrow_{CWT}$ NULL		NULL	
	$\Downarrow_A$ Disconnect	≪forward msg≫	NULL	
	$\Uparrow_N$ Release Timeout	≪forward msg≫	NULL	
DECISION	$\Downarrow_A$ Switch Calls	Hold$\Downarrow_A$	HELDCALL	
	$\Rightarrow_{CWT}$ NULL		NULL	
	$\Downarrow_A$ Disconnect	≪forward msg≫	NULL	
	$\Uparrow_N$ Release Timeout	≪forward msg≫	NULL	
HELDCALL	$\Rightarrow_{CWT}$ HELDCALL	Release Hold$\Downarrow_A$	ACTIVE	
	$\Rightarrow_{CWT}$ NULL	Alert$\Uparrow_A$	RINGBACK	
	$\Uparrow_N$ Release Timeout	≪forward msg≫	NULL	
ACTIVE	$\Downarrow_A$ Switch Calls	Hold$\Downarrow_A$	HELDCALL	
	$\Rightarrow_{CWT}$ NULL		NULL	
	$\Downarrow_A$ Disconnect	≪forward msg≫	NULL	
	$\Uparrow_N$ Release Timeout	≪forward msg≫	NULL	
RINGBACK	$\Downarrow_A$ Call Answered	Release Hold$\Downarrow_A$	NULL	
	▷Ringing Timeout	Disconnect$\Downarrow_A$	NULL	
	$\Uparrow_N$ Release Timeout	≪forward msg≫	NULL	

Table 1: Call Waiting Specification for Active Originating Call.

The transition labels in the feature's graphical specifications are mnemonic names for the events that activate transitions in the state-transition machines. Labels prefixed with symbol $\Rightarrow_{CWT}$ indicate that the transition is activated by another transition in one of the feature's other machines running concurrently on a different call stack. For example in the *Call Waiting* machines for both the new call and the active call, the transitions into HELDCALL are activated by an event *Switch Calls* indicating the the user is switching from one call to another; the transitions into ACTIVE are activated by a transition into HELDCALL in the parallel machine.

2.3 Tabular Specification Notation

Although graphical specifications provide an intuitive understanding of a feature's functional behavior, they are missing a lot of the detail needed to detect interactions. For example, the transitions have labels representing the event which causes the state transition, but there is no distinction between different types of events and there is no indication that the transitions have side-effects (e.g., that tokens are consumed or produced). We have designed a tabular notation for specifying a feature's functional behavior that is similar to the SCR tabular notation [8]. We intend feature specifications to be composed of both an intuitive graphical specification and a more precise tabular specification.

Tables 1 and 2 formally specify the behavior of the *Call Waiting* feature for an active call (Originating Call Model) and an incoming call, respectively. A third table specifying the behavior of the feature with respect to an active call (Terminating Call Model) is not shown; it would be the same as Table 1, with all occurrences of input $\Downarrow_N$ *Call Cleared* replaced by input $\Uparrow_N$ *Release Timeout*. Appendix A contains the tabular specifications of the Originating and Terminating Call Models.

State	Input	Output	NewState	Assertion
NULL	HUNTINGFACILITY⇒$_{TCM}$EXCEPTION	HUNTINGFACILITY⇒$_{TCM}$PRESENTINGCALL	HUNTINGFACILITY	⊨$uses\ bridge$
HUNTINGFACILITY	▷Facility Found		CALLSETUP	
	▷Busy	HUNTINGFACILITY⇒$_{TCM}$EXCEPTION	NULL	!$uses\ bridge$
	⇒$_{CWT}$NULL	HUNTINGFACILITY⇒$_{TCM}$NULL	NULL	!$uses\ bridge$
	⇓$_N$Call Cleared	HUNTINGFACILITY⇒$_{TCM}$NULL	NULL	!$uses\ bridge$
CALLSETUP	⇑$_A$Alert	Beep⇑$_A$	WAIT	
	⇒$_{CWT}$NULL		NULL	!$uses\ bridge$
	⇓$_N$Call Cleared	«forward msg»	NULL	!$uses\ bridge$
WAIT	⇒$_{CWT}$HELDCALL	Call Answered⇓$_A$	ACTIVE	
	⇒$_{CWT}$NULL		NULL	!$uses\ bridge$
	⇓$_N$Call Cleared	«forward msg»	NULL	!$uses\ bridge$
ACTIVE	⇓$_A$Switch Calls	Hold⇓$_A$	HELDCALL	
	⇒$_{CWT}$NULL		NULL	!$uses\ bridge$
	⇓$_A$Disconnect	«forward msg»	NULL	!$uses\ bridge$
	⇓$_N$Call Cleared	«forward msg»	NULL	!$uses\ bridge$
HELDCALL	⇒$_{CWT}$HELDCALL	Release Hold⇓$_A$	ACTIVE	
	⇒$_{CWT}$NULL	Alert⇑$_A$	RINGBACK	
	⇓$_N$Call Cleared	«forward msg»	NULL	!$uses\ bridge$
RINGBACK	⇓$_A$Call Answered	Release Hold⇓$_A$	NULL	!$uses\ bridge$
	▷Ringing Timeout	Disconnect⇓$_A$	NULL	!$uses\ bridge$
	⇓$_N$Call Cleared	«forward msg»	NULL	!$uses\ bridge$

Table 2: Call Waiting Specification for Incoming (Terminating) Call.

The tables specify the behavior of the feature in terms of functions and assertions. Each row in a table specifies a mapping from a state and an input event to a new state, a set of output events, and a set of raised assertions. Let T be the set of all tokens that the basic call models and features can generate and output to the call stack[2]; let inT be the set of input events denoting the receipt of a token $t \in T$ and let $outT$ be the set of events denoting the output of a token $t \in T$. Let N be the set of notifications of all state-transitions specified in the call models and features, and let inN and $outN$ be the sets of events denoting the receipt or the output of a notification, respectively. For a given feature f, let S be the feature's set of states, let I be the feature's set of internal signals, and let A be the set of assertions raised by the feature. Then formally, a tabular specification $\mathcal{T}$ represents the feature's transition relation: a partial function from states and input events to states, sets of output events, and sets of assertions.

$$\mathcal{T} : (S \times (inT \cup inN \cup I)) \mapsto (S \times \mathcal{P}(outT \cup outN) \times \mathcal{P}(A))$$

Figure 5 lists the type of input events, output events, and assertions that may appear in a tabular specification. We would like to emphasize that this is a preliminary list of events and assertions, and we expect that the list will grow as we gain more experience specifying features. Input events include tokens from the environment to the call model[3], tokens from the call model to the environment, *Feature Signals* between the feature's sub-machines on different call stacks, *Transition Attempt Notification* tokens from the underlying call model or from lower-level features, and local *Internal Signals* indicating the termination of internal processing. Output events include tokens that are output to either the call model or the environment and notifications of imposed state-transitions in the call model or in lower-level features. Assertions include established connections between agents, acquisitions of call-processing resources, dispossessions of resources, assertions on the values of tokens, and required relationships among more primitive assertions.

Note that there are two methods by which a feature can affect the operation of the underlying call model. One method is to generate tokens on behalf of the agent or the communications network (i.e., on behalf of the remote agent associated with an established connection); these tokens are passed down the call stack towards the call model and are expected to cause the call model to change state. For example in the RINGBACK state of *Call Waiting* in Table 1, the feature has alerted the user that he left a call on hold when he hung up the phone; if the user does not answer the ringback before the ▷Ringing Timeout event, then the feature terminates the held call by sending a *Disconnect*⇓$_A$ token to the call model indicating that the user has hung up.

The second method is to circumvent certain state-transitions in the call model or in lower-level features and replace them with new state-transitions. Notifications of all state-transitions are passed down the call stack so that features can act on them. If a feature receives notification of transition S⇒T, it can impose a new transition S⇒U *only if* there exists a transition from S to U in the lower-level machine that was making transition S⇒T. For example in Table 2, *Call Waiting* is initially activated by the transition from HUNTINGFACILITY to EXCEPTION in the Terminating Call Model (i.e., the feature is activated because the call model has determined that the user's line is busy). The *Call Waiting* feature consumes this state-transition notification, performs

---

[2]Note that set T will grow as new tokens are needed to implement new features.

[3]To enhance readability, tokens from the environment that are being passed down the call stack are annotated with their source, which is either the agent associated with call stack or the communications network. Tokens from features or from the call model to the environment are annotated with their destination.

Input events:

$\Downarrow_A\, token$ — information (interpreted as being from the agent) that is being passed down the call stack.

$\Downarrow_N\, token$ — information (interpreted as being from the call stack of connection's remote agent via the communications network) that is being passed down the call stack.

$\Uparrow_A\, token$ — information that is being passed up the call stack towards the agent.

$\Uparrow_N\, token$ — information that is being passed up the call stack towards the communications network (i.e., towards the call stack of the connection's remote agent).

$\Rightarrow_f S$ — a signal from one of the feature's other machines, indicating that it has transitioned into state S. This signal is only visible to other state-transition machines associated with the same instantiation of the feature.

$S1 \Rightarrow_{CM} S2$ — a signal that the underlying call model is transitioning from S1 to S2. This signal is passed down the call stack so that the activated features (in the order of their priority) have the opportunity to circumvent the transition and impose their own desired transition in the underlying call model. Imposed call model transitions are treated as original call model transitions and are passed down the call stack (again) so that the activated features have the opportunity to circumvent the new imposed transition.

$S1 \Rightarrow_f S2$ — a signal that an underlying feature f is transitioning from S1 to S2. As with call model transitions, feature transitions are passed down the call stack ending at the affected feature, so that higher-priority features can circumvent the transition and impose their own desired transition.

$\triangleright event$ — a signal indicating the termination of internal processing. This signal is only visible to the machine performing the internal processing.

Output events:

$token\Downarrow_A$ — a *token* is sent from the feature to the call model, which lower-level machines will assume is from the agent.

$token\Downarrow_N$ — a *token* is sent from the feature to the call model, which lower-level machines will assume is from the communications network (i.e., from the call stack of the connection's remote agent).

$token\Uparrow_A$ — a *token* is sent from the feature to the agent, which will appear to be from the call model.

$token\Uparrow_N$ — a *token* is sent from the feature to the communications network (i.e., the call stack of the connection's remote agent), which will appear to be from the call model.

$\ll forward\ msg \gg$ — the input token is forwarded to the next level machine without alteration.

$S1 \Rightarrow_{CM} S2$ — the feature imposes a new state transition from S1 to S2 in the call model.

$S1 \Rightarrow_f S2$ — the feature imposes a new state transition from S1 to S2 in lower-level feature f.

NewCall(CM) — instantiation of a new call stack, based on either an Originating Call Model or a Terminating Call Model.

Assertions:

$connect(A,B)$ — a connection is established between agents **A** and **B**.

$uses\ X$ — the feature acquires resource X.

$Q(token)$ — the feature asserts Q on the value of *token*.

$\models A$ — the feature raises assertion A, which continues to hold along all computation paths until it is lowered. A is a propositional logic formula, where the propositions are primitive assertions (e.g., $Q(token)$).

$!A$ — the feature lowers assertion A.

Figure 5: Notation used in tabular specifications.

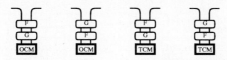

Figure 6: Call configurations: features F and G invoked by same agent.

Figure 7: Call configurations: features F and G invoked by different agents.

it own HUNTINGFACILITY operation using the new information that *Call Waiting* has been activated, and outputs a new call model transition based on the results of its own HUNTINGFACILITY computation.

3 Algorithm for Detecting Feature Interactions

Feature specifications are analyzed by composing feature machines and tracing the paths through the composite machine. There is a trace path for each possible call processing sequence. Feature interactions can be detected by testing the reachable states of the composite machine. Some interactions are detected as a conflict between the information required in a state and the information actually present in that state, some are detected as a relationship between assertions in a particular state of the composite machine, and some are detected as a perturbation in the paths through a component machine when it is part of the composite machine.

In this section we illustrate how the feature machines are composed and how the composite machines are traced. In addition, we introduce four categories of interactions and show how traces through the composite machine can be used to detect them.

3.1 Composing Feature Machines

The composite machine for a set of features is built by executing all possible configurations of the component feature machines on all possible call models. Suppose we want to determine whether or not features F and G interact. We would need to compose the behaviors of the two features with respect to all possible call configurations that involve both features. If the features can be invoked by the same agent, then all possible arrangements of the two features on the agent's call stack need to be traced. For example, if features F and G can be invoked on either an Originating Call Model or a Terminating Call Model, then we would need to search four call configurations as shown in Figure 6. If the features can be invoked by different agents (A and B) involved in the same telephone call, then there are two additional configurations that need to be traced, as shown in Figure 7

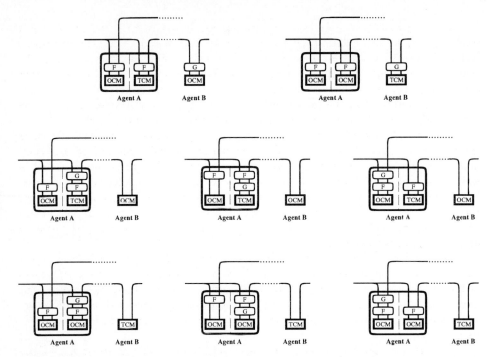

Figure 8: Call configurations: feature F invokes new Originating Call stack on behalf of agent A.

Suppose that F adds a new call to the configuration, thereby adding a new call stack to agent A's original call stack. Then the composite machine consists of all possible combinations of *three* call stacks: each of agent B's allowable call stacks must be composed with the allowable pairs of A's call stacks. Assuming that F creates a new Originating Call Model on behalf of agent A, then there are two configurations in which the features are invoked by separate agents, as shown in the top row of Figure 8; and there are six configurations in which the features are invoked by the same agent, as shown in the second and third rows of Figure 8[4] All eight configurations need to be traced.

In general, the complexity of the composite machine will depend on whether the features operate on the same call stack, whether the features reside with the same user, whether the features can add new call stacks to the configuration, and whether the features can be invoked on Originating Call Models, Terminating Call Models, or both. All possible call configurations must be subjected to the tracing algorithm.

3.2 Tracing the Composite Machine

The reachability graph for the composition of two features can be built using a backtracking algorithm. Consider a trace on a call from agent A to agent B[5]. The trace starts with the OCM for A. If A's OCM proceeds to the point where a call request is

---

[4] In Figure 8, the new call stack created by feature F is the leftmost call stack for agent A.

[5] This call may also involve Originating or Terminating Calls from other agents depending on the features that A and B invoke.

sent up to agent B then a TCM for B commences. If at any time an event in a call stack triggers a feature, then that feature is placed on the stack in the appropriate state. If an event causes a transition in a feature then that transition occurs in the composite machine. Whenever a point of non-determinism is reached (where either one machine can make different transitions or where several machines can transition in different orders or on the same event), then backtracking is used to follow all the possible trace paths. The trace ends when both features have deactivated.

Some possible paths in the composite machine, or in a feature, can have arbitrary length due to cycles. This is especially true of features which respond to user commands. Consider an agent using *Three Way Calling* to alternately phone A and B while maintaining a call with C. If the feature's behavior does not depend on the number of times it traverses the cycle then it suffices to trace the cycle once. Hence for these features we need only trace acyclic or uni-cyclic paths and these are all of finite length. We have not yet discovered a feature whose behavior depends on the number of times a cycle is traversed.

The complexity of the tracing will depend on the complexity of each feature, on the amount of synchronization between the features, on whether they are in the call stacks of the same or different agents, and on how long the features' lifespans overlap. The number of paths will generally be small relative to the combinatorial possibilities because each feature will synchronize with its underlying call model and progress towards its deactivation.

Although the search space can explode exponentially in theory, we do not believe this makes exhaustive search infeasible. The invocation requirements of a feature often reduces the number of call stack configurations that are possible. Features represented by multiple machines executing in parallel usually have synchronized transitions. In addition most features can be represented by machines with few states, and most features have little branching and proceed almost linearly. *Call Waiting* for instance is a complex feature, but the machine for each stack has only six states, and the graph is of low degree. As a result, the reachability graph of each configuration is relatively small and can quickly be traced.

3.3 Detecting Feature Interactions

All the kinds of interactions discussed in this paper can be detected by testing states as opposed to paths in a composite machine's reachability graph. This means that with careful bookkeeping they can all be found by tests made while tracing the reachability graph.

Our tracing algorithm analyzes requirements and design specifications, so it can only be used to detect interactions that manifest themselves at the specification stage; implementation-dependent interactions and race conditions will not be detected. In addition, we have concentrated our efforts on detecting interactions among call processing features; we have not addressed interactions among features associated with operations, administration, and maintenance services, such as billing. The classes of interactions we have been able to detect include *call control interactions, resource contentions, information invalidations,* and *assertion invalidations.*

3.3.1 Call Control Interactions

Some interactions arise because different features try to interpret the same signals simultaneously with conflicting results. In our model these *call control interactions* show up when the current states of different features on the same call stack each have a transition with the same token as an input event coming down the call stack. When this happens only one feature may receive the signal, thereby changing the expected behavior of the other.

Call control interactions are found by comparing the active states of the features on all call stacks for a single agent. The comparison is made on the input events of the transitions exiting the active states. If the same event occurs in the input portion of any two transitions then a call control interaction is detected: both features are vying for the same event at the same time as it passes down the call stacks.

3.3.2 Resource Contention

Resource contentions arise when features try to share an unsharable resource R. If the feature G tries to acquire R after the feature F has already done so and before F has released R, then there is a resource contention. Currently we restrict our attention to resource contentions among features of a single agent. In our model a feature indicates that it needs to use a resource R by raising the assertion *uses R*. When the feature is finished with the resource it raises *!uses R*.

When a feature asserts *uses R*, the assertion is added to the assertion list of that agent. When *!uses R* is raised *uses R* is removed from the assertion list. Finding resource contentions involves checking the list of assertions for each agent. If one feature tries to obtain resource R while another is using it then both features will raise *uses R*. A duplicate insertion of *uses R* in the assertion list of the same agent indicates a resource contention. There is an obvious generalization for sharing n copies of a resource.

3.3.3 Information Invalidation

Recall that in our system information flowing up and down call stacks is represented as tokens. An *information invalidation* arises when a feature alters information which a another feature subsequently uses for a decision. If the alteration is the insertion or removal of an information token, then the sequence of tokens the second feature receives may perturb the reachability graph of the second feature. These interactions are detected by comparing the reachability graph of the second feature viewed on its own with its reachability graph when viewed as a component of the composite machine. This type of analysis requires an algorithm that tests for graph inclusion. In this paper, we are only addressing analysis tests that require a linear trace through the composite machine's reachability graph; thus, we will not discuss this type of information invalidation, which we call *restriction invalidation*, any further.

If the alteration of information consists of a change in the value but not the name of a token (such as changing the directory number of the caller or callee), then this is represented in the model by appending a prime (') to the token. Interactions caused by this kind of alteration are detected when the primed token (*token'*) arrives in a state of the second feature that is expecting to receive *token*. In our model a state uses the information stored in a token if and only if that token is named in the input, output, or assertion portion of a transition leaving the state. Naming the token on a transition means that the feature will make a decision or take an action using the value extracted from that token as it passes through the state.

Information invalidations are detected when a primed token t' arrives in a state of a feature machine *and* that state has a transition which names token t in its input, output, or assertion portion. Since the state has received primed token t', the information that the feature needs has been changed, indicating an information invalidation. Note that such interactions may be intended. Feature F may be explicitly designed to interact with feature G via changed tokens. We want to be able to detect these interactions to help verify that they occur when they should.

3.3.4 Assertion Invalidation

In an information invalidation, a feature's assumption about a piece of information is invalidated by a previous feature. The converse is also possible: a feature may make an assumption that is invalidated by a subsequent feature. We call this an *assertion invalidation*. A feature expresses its expectation about the future (e.g., the future of a token or the future of the telephone connection) by raising assertions and raising assertion requirements. An assertion is a predicate. An assertion requirement is a logic relation that must hold on assertions. An example requirement is $\models(P \rightarrow Q)$, which means that whenever assertion P has been raised, assertion Q must also have been raised[6].

Assertion invalidations are manifested when an assertion requirement fails on a path in the composition. Assertion invalidations are detected by testing assertions and assertion requirements in each state. The set of assertions and assertion requirements in a state is the union of the assertion lists of all the agents in the call. For example, an assertion requirement might be $\models(connect(A,B) \rightarrow X(A))$. This can be tested by examining the assertion lists for assertions $X(A)$ and $connect(A,B)$ to determine if $X(A)$ is asserted whenever $connect(A,B)$ is asserted.

4 Case Studies

We present four examples of feature interactions that can be detected using our analysis technique.

4.1 Information Interaction

This example presents a fairly complete informal trace of the interaction of two features: *Calling Number Display* and *Calling Number Display Blocking*. *Calling Number Display* presents the directory number of the caller to the callee when the phone rings. *Calling Number Display Blocking* prevents the caller's directory number from being presented to the callee. The (intended) interaction between these features is that *Calling Number Display Blocking* will prevent *Calling Number Display* from properly displaying the caller's number.

The feature machine for *Calling Number Display* has one active state (see Figure 9). It records the originating caller's number when that token passes through the state, and presents the directory number to the user when the phone is rung. The feature machine for *Calling Number Display Blocking* also has only one active state (see Figure 10). This state intercepts the token which indicates the caller's number and 'changes' it such that it can no longer be displayed.

Each feature operates on only one call model: *Calling Number Display Blocking* works on an Originating Call Model (OCM) and *Calling Number Display* works on a Terminating Call Model (TCM). Thus there is only one configuration of the call stacks

---

[6] Note that the meaning of the assertion requirement is NOT that the truth of assertion Q can be inferred from the truth of assertion P.

State	Input	Output	NewState
NULL	$\Downarrow_N$ Termination Attempt	≪forward msg≫	WAITFORRING
WAITFORRING	$\Uparrow_A$ Alert	≪forward msg≫ Termination Attempt.origin$\Uparrow_A$	NULL
	$\Downarrow_A$ Disconnect	≪forward msg≫	NULL
	$\Downarrow_N$ Call Cleared	≪forward msg≫	NULL

Figure 9: Specification of the *Calling Number Display* feature.

State	Input	Output	NewState
NULL	SELECTINGROUTE▷$_{OCM}$AUTHCALLSETUP	≪forward msg≫	WAITFORSEND
WAITFORSEND	$\Uparrow_N$ Call Request	Call Request.origin'$\Uparrow_N$	NULL
	$\Downarrow_A$ Disconnect	≪forward msg≫	NULL
	$\Uparrow_N$ Release Timeout	≪forward msg≫	NULL

Figure 10: Specification of the *Calling Number Display Blocking* feature.

to trace.

Figure 11(a) shows the first stage of the trace. The composite machine consists of a single call model: the caller's OCM. Next the OCM begins to set up a call. When it reaches state AUTHCALLSETUP, the feature *Calling Number Display Blocking* is activated and placed on the call stack. Upon activating, *Calling Number Display Blocking* immediately transitions into the WAIT state . If the call setup is authorized, the OCM produces the token $\Uparrow_N$ *CallRequest* which is sent towards the destination agent B to initiate a call. *CallRequest* is a compound token with several fields; one of these fields is *origin*, which represents the caller's directory number (see Figure 11(b)). Token $\Uparrow_N$ *Call Request* passes up the call stack to *Calling Number Display Blocking*. *Calling Number Display Blocking* forwards the message up the call stack with the *origin* marked as modified[7]. The modification is shown by appending a prime (') to *origin*. When the (modified) *Call Request* token is passed upwards, *Calling Number Display Blocking* transitions to NULL.

The *Call Request* token arrives at the destination agent as a *Termination Attempt* token. The arrival of this token activates both a Terminating Call Model for the agent

---

[7]In the real world implementation of this feature, it is actually a permission on the origin information that is modified.

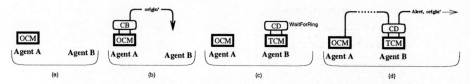

Figure 11: Example of an interaction due to information invalidation.

and the *Calling Number Display* feature (see Figure 11(c)). As it activates *Calling Number Display*, the feature transitions into the state WAITFORRING. One of the transitions from this state outputs the token *origin* (which is obtained from the compound token *Termination Attempt*). As the *Termination Attempt* token passes through state WAITFORRING, the presence of the primed token *origin'* is noted and the information invalidation is detected at this point. Intuitively, *Calling Number Display* is attempting to record for future use information that has been changed[8]. *Calling Number Display* passes the *Termination Attempt* token down towards the underlying call model. When the TCM reaches the state ALERTING, it sends an *Alert* token up to the agent. When this token reaches *Calling Number Display*, it is forwarded up the call stack to the agent. In addition, a new token *Termination Attempt.origin* is created and it also sent up the call stack to the agent (see Figure 11(d)). This new token contains the directory number to be displayed (which, due to the interaction, has been zeroed out). Finally, *Calling Number Display* transitions to NULL and the trace is complete.

No other interactions are found. No assertions were raised, so no assertion invalidations or resource contentions are detected. In addition, there are no call control interactions since no states of the two feature machines were ever active on the same call stack.

4.2 Call Control Interaction

This example demonstrates the detection of a call control interaction between *Call Waiting* and *3-Way Calling*. *3-Way Calling* allows an agent who is engaged on a line to place that line on hold, receive a dial tone, and dial a third party. He may then speak privately with that third party. If he wishes, he may end the connection to the third party and return to the held call, or he may add the third party to the original conversation making it a 3-Way call. The specification for *Call Waiting* was presented earlier in the paper; the specification for *3-Way Calling* appears in Figures 12 and 13.

The agent indicates his choices by initiating an event. Each possible event can be mapped to one of several input signals [11, 15]. On a simple telephone the events to initiate the second call and to merge it with the first might be mapped to the flashhook. Similarly, the signal to accept an incoming call when *Call Waiting* has been activated might be mapped to the flashhook. In this case the features will interact.

For this example we will not produce a full trace, but simply exhibit the state of the compound machine where the interaction occurs. This state can be reached by the following series of events. Agent A, who subscribes to both features, is in a call with B. Agent A decides to call party C and uses *3-Way Calling* to do so[9]. While A is

---

[8] While our algorithm detects the interaction at this point, the effects of the interaction will not be apparent to the agent yet.

[9] In reality, this situation cannot in fact occur due to the resource contention between these features which we describe in Example 4.

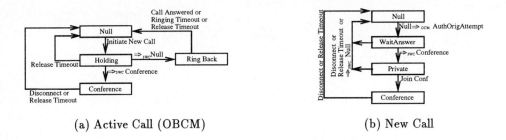

(a) Active Call (OBCM) (b) New Call

State	Input	Output	NewState	Assertion
NULL	$\Downarrow_A$ Initiate New Call	NewCall(OCM) $Hold\Downarrow_A$	HOLDING	$\models$ uses bridge
HOLDING	$\Rightarrow_{3WC}$ CONFERENCE	Release Hold$\Downarrow_A$	CONFERENCE	
	$\Rightarrow_{3WC}$ NULL	Alert$\Uparrow_A$	RINGBACK	
	$\Uparrow_N$ Release Timeout	$\ll$ forward msg $\gg$	NULL	!uses bridge
CONFERENCE	$\Downarrow_A$ Disconnect	$\ll$ forward msg $\gg$	NULL	!uses bridge
	$\Uparrow_N$ Release Timeout	$\ll$ forward msg $\gg$	NULL	!uses bridge
RINGBACK	$\Downarrow_A$ Answered	Release Hold$\Downarrow_A$	NULL	!uses bridge
	$\triangleright$ Ringing Timeout	Disconnect$\Downarrow_A$	NULL	!uses bridge
	$\Uparrow_N$ Release Timeout	$\ll$ forward msg $\gg$	NULL	!uses bridge

Figure 12: Specification of *Three-Way Calling* on an active call (OCM).

State	Input	Output	NewState	Assertion
NULL	NULL $\Rightarrow_{OCM}$ AUTHORIGATTEMPT	$\ll$ forward msg $\gg$	WAITANSWER	
WAITANSWER	$\Downarrow_A$ Answered	$\ll$ forward msg $\gg$	PRIVATE	
	$\Rightarrow_{3WC}$ NULL		NULL	
	$\Downarrow_A$ Disconnect	$\ll$ forward msg $\gg$	NULL	
	$\Uparrow_N$ Release Timeout	$\ll$ forward msg $\gg$	NULL	
PRIVATE	$\Downarrow_A$ Join Conf		CONFERENCE	
	$\Rightarrow_{3WC}$ NULL		NULL	
	$\Downarrow_A$ Disconnect	$\ll$ forward msg $\gg$	NULL	
	$\Uparrow_N$ Release Timeout	$\ll$ forward msg $\gg$	NULL	
CONFERENCE	$\Downarrow_A$ Disconnect	$\ll$ forward msg $\gg$	NULL	
	$\Uparrow_N$ Release Timeout	$\ll$ forward msg $\gg$	NULL	

Figure 13: Specification of *Three-Way Calling* on the new call.

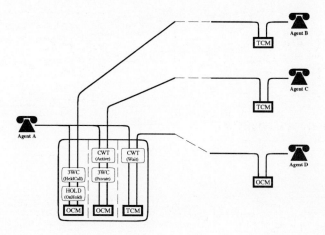

Figure 14: Example of an interaction due to call control interaction.

engaged in a private conversation with C, he receives a call from party D, and accepts it using *Call Waiting*. The state of the composite machine is shown in Figure 14[10]. In this composite state, both state ACTIVE in *Call Waiting* and state PRIVATE in *3-Way Calling* have transitions labeled $\Downarrow_A$ *flashhook*, indicating a call control interaction.

Note that it is not enough to simply check if both features accept the same input event coming down the stack. Two features can both accept and act on the same event without an interaction if in the features' composite machine, the features are never ready to accept the event at the same time.

4.3 Assertion Interaction

Originating Call Screening attempts to prevent a connection originating from the agent to any one of a specified set of directory numbers. *Call Forwarding* attempts, under conditions that vary from one version of the feature to another, to connect an incoming call to a new target. *Call Forwarding* can interfere with *Originating Call Screening*. If agent A has placed agent C's number in his *Originating Call Screening* screening list but has not placed agent B's number in the list, and if calls to B have been forwarded to C's number, then A may reach C by dialing B.

The specification for feature *Originating Call Screening* is shown in Figure 15. Certain transitions in the *Originating Call Screening* machine are labeled with $\models OCS(target)$. This indicates that the *target* (directory number) passed the test *Originating Call Screening* applied. $OCS(target)$ can be thought of as a label applied to successful paths through the *Originating Call Screening* machine. *Originating Call Screening* also asserts requirement $\models connect(A, ?X) \rightarrow OCS(?X)$, which states that along any path resulting in a connection originating at A and terminating at X, assertion $OCS(X)$ must hold[11]. The assertion $\models connect(A,B)$ is raised by the tracing algorithm when a connection is made by the underlying call models; alternatively, it may be raised by another feature that implements a virtual connection between the agents. This state of the composite machine is shown at the top of Figure 16.

---

[10]When *3-Way Calling* is used to initiate a new call, the *3-Way Calling* places the active line on hold by invoking the feature *Hold* on the call stack of the active line.

[11]In the above assertion, A is the agent that invokes the *Originating Call Screening* feature, and X is a free variable.

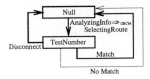

State	Input	Output	NewState	Assertion
NULL	$\Downarrow_N$ AnalInfo$\Rightarrow_{OCM}$ SelectRoute		TEST	
TEST	$\triangleright$ Match	AnalInfo$\Rightarrow_{OCM}$ Exception	NULL	$\models(\neg OCS(target))$
	$\triangleright$ NoMatch	AnalInfo$\Rightarrow_{OCM}$ SelectRoute	NULL	$\models(connect(A,B) \rightarrow$ $OCS(B))$ $\models OCS(target)$
	$\Downarrow_A$ Disconnect	$\ll$ forward msg $\gg$	NULL	
	$\Uparrow_N$ Release Timeout	$\ll$ forward msg $\gg$	NULL	

Figure 15: Specification of the *Originating Call Screening* feature.

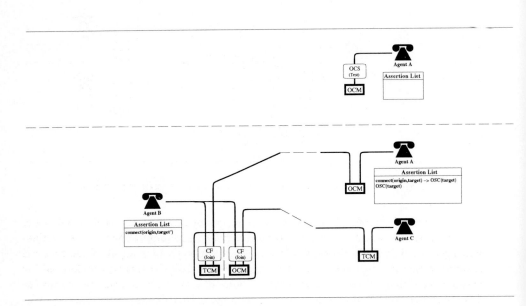

Figure 16: Example of an interaction due to an assertion invalidation.

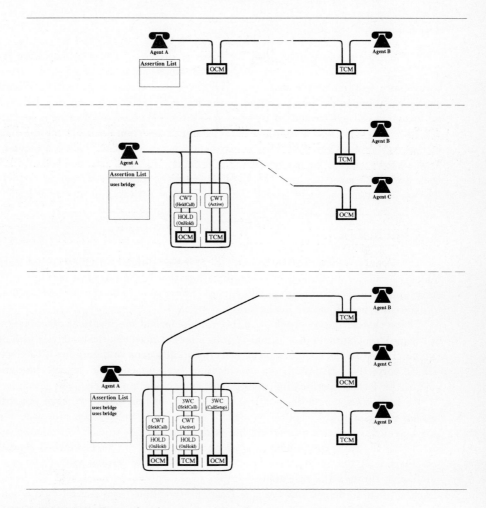

Figure 17: Example of an interaction due to resource contention.

When *Call Forwarding* is activated on an incoming call, *Call Forwarding* initiates a new call from the agent (who is the terminus of the incoming call) to a new destination agent, and forms a virtual connection from the originating agent of the first call (*origin*) to the new destination agent (*target'*). When *Call Forwarding* has done all the preliminary signaling necessary to form the virtual connection, it transitions into the JOIN state and raises the assertion $\models connect(origin, target')$.

The bottom of Figure 16 shows the interactions of these two features when agent A invokes *Originating Call Screening* and agent B forwards all of his calls to agent C. In this state of the composite machine, assertion *connect(origin,target')* has been raised but the required assertion *OCS(target')* has not. Note that the interaction occurs after *Originating Call Screening* has deactivated. The assertions raised by *Originating Call Screening* server to record invariants expected to hold throughout the call – even after the asserting feature has been deactivated.

4.4 Resource Interaction

3-Way Calling and *Call Waiting* both require the use of a piece of hardware known as a bridge, but there is only one bridge available to an agent. If an agent attempts to use both features at the same time there will be a resource contention. Figure 17 shows an abbreviated trace. Initially, the agent is involved in a call. He receives an incoming call and accepts it using *Call Waiting*. When *Call Waiting* is invoked, it asserts *uses bridge*. This is shown in the subscriber's assertion list in the middle diagram in Figure 17. When the subscriber subsequently invokes *3-Way Calling*, a second *uses bridge* is asserted (see the bottom diagram in Figure 17). This assertion will also go into the subscriber's assertion list. This duplicate insertion indicates a resource contention. The same contention would arise if the features were activated in reverse order.

5 Conclusion

The techniques described in [9] and [13] propose template specifications of features. One of the major advantages of a template notation is the power of its expressibility. As a result, the specifications contain enough information to be able to detect a number of different types of feature interactions.

The work described in [1], [5], and [14] propose formal specifications of features. One of the major advantages of a formal approach is that mathematical analysis techniques can be applied to the composition of feature specifications to determine if the features behave correctly. [1] describes a specification environment that provides automated support for formal specification, refinement, and simulation of telephone features.

We have taken a middle-of-the-road approach. In this paper, we have presented a tabular notation that is flexible enough to support the specification of a wide variety of telephony events and properties, but is rigorous enough to allow automated analysis. Using the call stack model of a telephone call [12], we have sketched algorithms for determining the set of composite machines associated with a pair of features and for tracing through the reachability graphs of the composite machines. We have also described how to detect four types of feature interactions (call control interaction, information invalidation, resource contention, and assertion invalidation) by examining the information known at each state of the composite machines' reachability graphs. Our next goal is to automate the algorithms presented in this paper.

References

[1] R. Boumezbeur and L. Logrippo. "Specifying Telephone Systems in LOTOS". *IEEE Communications*, 31(8):38–45, August 1993.

[2] R. Brooks. "A Robust Layered Control System for a Mobile Robot". *IEEE Journal of Robotics and Automation*, RA-2:14–23, April 1986.

[3] E.J. Cameron, N. Griffeth, Y. Lin, and H. Velthuijsen. "Definitions of Services, Features, and Feature Interactions", December 1992. Bellcore Memorandum for Discussion, presented at the International Workshop on Feature Interactions in Telecommunications Software Systems.

[4] E.J. Cameron, N. Griffeth, Y.J. Lin, M. Nilson, W. Schnure, and H. Velthuijsen. "A Feature Interaction Benchmark in IN and Beyond". Technical Report TM-TSV-021982, Network Systems Specifications Research, Bell Communications Research, September 1992.

[5] A. Fekete. "Formal Models of Communication Services: A Case Study". *IEEE Computer*, 26(8):37–47, August 1993.

[6] A. Flynn, R. Brooks, and L. Tavrow. "Twilight Zones and Cornerstones: A Gnat Robot Double Feature". Technical Report A.I. Memo 1126, Artificial Intelligence Laboratory, Massachusetts Institute of Technology, 1989.

[7] N. Griffeth and Y. Lin. "Extending Telecommuncations Systems: The Feature-Interaction Problem". *IEEE Computer*, 26(8):14–18, August 1993.

[8] K. Heninger. "Specifying Software Requirements for Complex Systems: New Techniques and Their Applications". *IEEE Transactions on Software Engineering*, SE-6(1):2–12, January 1980.

[9] E. Kuisch, R. Janmaat, H. Mulder, and I. Keesmaat. "A Practical Approach to Service Interactions". *IEEE Communications*, 31(8):24–31, August 1993.

[10] Northern Telecom. *DMS-100 Meridian Digital Centrex Library*, 50039.08/12-92 issue 1 edition, 1992.

[11] D. Parnas and J. Madey. Functional Documentation for Computer Systems Engineering (Version 2). Technical Report CRL Report 237, Department of Electrical and Computer Engineering, McMaster University, 1991.

[12] G. Utas. "Feature Processing Environment", December 1992. Presented at the International Workshop on Feature Interactions in Telecommunications Software Systems.

[13] Y. Wakahara, M. Fujioka, H. Kikuta, and H. Yagi. "A Method for Detecting Service Interactions". *IEEE Communications*, 31(8):32–37, August 1993.

[14] P. Zave. "Feature Interactions and Formal Specifications in Telecommunications". *IEEE Computer*, 26(8):20–30, August 1993.

[15] P. Zave and M. Jackson. "Conjunction as Composition". *ACM Transactions on Software Engineering and Methodology*, 2(4):379–411, October 1993.

Appendix A: Specification of Call Model

Figures 18 and 19 contain the tabular specifications of the originating basic call model (OBCM) and the terminating basic call model (TBCM), respectively.

State	Input	Output	NewState
NULL	$\Downarrow_A$ Origination Attempt		AUTHORIGATTEMPT
AUTHORIG ATTEMPT	$\triangleright$ Originated	Collect Info$\Uparrow_A$	COLLECTINGINFO
	$\triangleright$ Origination Denied	Origination Denied$\Uparrow_A$	EXCEPTION
	$\Downarrow_A$ Disconnect		NULL
COLLECTING INFO	$\Downarrow_A$ Info Collected		ANALYZINGINFO
	$\triangleright$ Collection Timeout	Collection Timeout$\Uparrow_A$	EXCEPTION
	$\Downarrow_A$ Disconnect		NULL
ANALYZING INFO	$\triangleright$ Valid Info		SELECTINGROUTE
	$\triangleright$ Invalid Info	Invalid Info$\Uparrow_A$	EXCEPTION
	$\Downarrow_A$ Disconnect		NULL
SELECTING ROUTE	$\triangleright$ Route Selected		AUTHCALLSETUP
	$\triangleright$ Network Busy	Network Busy$\Uparrow_A$	EXCEPTION
	$\Downarrow_A$ Disconnect		NULL
AUTH CALLSETUP	$\triangleright$ Call Setup Authorized	Call Request$\Uparrow_N$	SENDCALL
	$\triangleright$ Call Setup Denied	Call Set Denied$\Uparrow_A$	EXCEPTION
	$\Downarrow_A$ Disconnect		NULL
SENDCALL	$\Downarrow_N$ Call Delivered	Call Delivered$\Uparrow_A$	ALERTING
	$\Downarrow_N$ Route Busy		SELECTINGROUTE
	$\Downarrow_N$ Answered	Answered$\Uparrow_A$	ACTIVE
	$\Downarrow_N$ Called Party Busy	Called Party Busy$\Uparrow_A$	EXCEPTION
	$\Downarrow_N$ Call Cleared	Call Cleared$\Uparrow_A$	EXCEPTION
	$\Downarrow_A$ Disconnect	Call Cleared$\Uparrow_N$	NULL
ALERTING	$\Downarrow_N$ Answered		ACTIVE
	$\Downarrow_N$ Called Party Busy		EXCEPTION
	$\Downarrow_N$ Call Cleared	Call Cleared$\Uparrow_A$	EXCEPTION
	$\Downarrow_A$ Disconnect	Call Cleared$\Uparrow_N$	NULL
ACTIVE	$\Downarrow_N$ Call Cleared		RELEASEPENDING
	$\Downarrow_A$ Disconnect	Call Cleared$\Uparrow_N$	NULL
RELEASE PENDING	$\Downarrow_N$ Called Party Reconnect		ACTIVE
	$\triangleright$ Release Timeout	Release Timeout$\Uparrow_N$	NULL
	$\Downarrow_A$ Disconnect	Call Cleared$\Uparrow_N$	NULL
EXCEPTION	$\Downarrow_A$ Disconnect		NULL

Figure 18: Specification of Basic Call Model for Originating Caller.

State	Input	Output	NewState
NULL	$\Downarrow_N$ Termination Attempt		AUTHTERMINATION
AUTH TERMINATION	$\triangleright$ Call Presented		HUNTINGFACILITY
	$\triangleright$ Termination Denied	Call Cleared$\Uparrow_N$	EXCEPTION
	$\Downarrow_N$ Call Cleared		NULL
	$\Downarrow_A$ Disconnect	Call Cleared$\Uparrow_N$	NULL
HUNTING FACILITY	$\triangleright$ Facility Found		PRESENTINGCALL
	$\triangleright$ Busy	Called Party Busy$\Uparrow_N$	EXCEPTION
	$\Downarrow_N$ Call Cleared		NULL
	$\Downarrow_A$ Disconnect	Call Cleared$\Uparrow_N$	NULL
PRESENTING CALL	$\triangleright$ Call Accepted	Call Delivered$\Uparrow_N$ Alert$\Uparrow_A$	ALERTING
	$\triangleright$ Call Failure		HUNTINGFACILITY
	$\triangleright$ Call Rejected	Call Cleared$\Uparrow_N$	EXCEPTION
	$\Downarrow_A$ Connected	Answered$\Uparrow_N$	ACTIVE
	$\Downarrow_N$ Call Cleared		NULL
ALERTING	$\Downarrow_A$ Connected	Answered$\Uparrow_N$	ACTIVE
	$\triangleright$ Call Rejected	Call Cleared$\Uparrow_N$	EXCEPTION
	$\triangleright$ Ringing Timeout	Call Cleared$\Uparrow_N$	EXCEPTION
	$\Downarrow_N$ Call Cleared		NULL
ACTIVE	$\Downarrow_A$ Disconnect	Call Cleared$\Uparrow_N$	RELEASEPENDING
	$\Downarrow_N$ Call Cleared		NULL
RELEASE PENDING	$\Downarrow_A$ Called Party Reconnect	Called Party Reconnect$\Uparrow_N$	ACTIVE
	$\Downarrow_N$ Release Timeout		NULL
	$\Downarrow_N$ Call Cleared		NULL
EXCEPTION			NULL

Figure 19: Specification of Basic Call Model for Terminating Called Party.

Classification, Detection and Resolution of Service Interactions in Telecommunication Services

Tadashi OHTA, Yoshio HARADA

ATR Communication Systems Research Laboratories
2-2, Hikari-dai, Seika-cho, Soraku-gun, Kyoto 619-02, Japan

Abstract. This paper proposes a way to classify, detect and resolve service interactions in telecommunication services. First, a general framework for classifying the interactions is described and then, based on our experiments, a detailed explanation of the classification, detection and resolution of interactions in the service specification design stage are described.

1. Introduction

The development of the information society has brought with it a demand for rapid and diverse telecommunication service provisioning on communication network systems. When many services are provided simultaneously, service conflicts such as those between service specifications, processing and execution can occur. These conflicts, called service interactions, can cause unexpected and undesirable results [1].

This paper proposes a way to classify, detect and resolve service interactions in telecommunication services. First, a general framework for classifying the interactions is described and then, based on our experiments, we give a detailed explanation of the classification, detection and resolution of interactions in the service specification design stage.

There are many types of service interactions, and no one method can resolve all of them [2]~[5]. To solve the problems, it is necessary to adopt a method that is appropriate to the type of interaction. Furthermore it is obviously desirable to resolve service interactions in the early stage of software development.

At ATR, we have developed a service specification description language called STR, which is based on the state transition rule. Methods of automatically detecting and resolving service interactions that use STR are being investigated at ATR. This paper is based on the results of our experimental investigation.

The paper is structured as follows. In section 2, we discuss where service interactions occur in the software development stage. In section 3, a detailed method of classifying and detecting service interactions in the service specification design stage is described. In section 4, support methods for resolving the interactions are described.

2. Classification according to Software Development Stage

It is very difficult to define a service interaction. But, to resolve service interactions, we must first define the interactions. Classifying service interactions is one of the ways to define the service interactions. In this section, classification regarding software development stage is described.

Based on where the service interactions are detected, the software development process is divided into three stages: service specification design, program manufacturing and program execution.

In the service specification design stage, service interactions are caused by the service specifications themselves, which are independent of the implementation of services. There are some conflicts between service specifications. This can be called service interaction narrowly defined.

In the program manufacturing stage, service interactions are caused by conflicts between program modules and between data both of which depend on actual implementation of service specifications. These conflicts are not stochastic but rather deterministic conflicts, which can be found by tracing the program.

In the program execution stage, service interactions are caused by dynamic conflicts between software and hardware resources. These conflicts are stochastic, and thus difficult to find in the earlier stages.

3. Interaction at the Service Specification Design Stage

In this section, a detailed classification of the service specification design stage is given based on the results of our experiments. First of all, because the service specification can be represented as a FSM(Finite State Machine), the interaction problem is described from the standpoint of the FSM. Then, the service model treated in this paper is introduced. The service specification description language developed by ATR is also briefly described. Finally, some examples of interactions are given.

3.1 Service Specification and FSM

Telecommunication services can be represented as an FSM (Finite State Machine). Therefore, the classification of service interaction problems, in the service specifying stage, can be treated from the following standpoints:

cl-1) Logical problems of the FSM:

cl-1-1) No transition: In the FSM, all states must have at least one succeeding transition, otherwise the transitions stop at a certain state. All states must return to the initial state, otherwise a loop occurs among the states. For telecommunication services, the initial state can be regarded as an idle state.

cl-1-2) Multiple transitions for the same event: A state must not have different next transitions when the same event occurs at the state, otherwise nondeterminacy occurs. The above problems are treated as logical problems of the FSM.

cl-2) Semantical problems of the FSM:

Except for the logical problems of the FSM, the following problems must be checked from the standpoints of the combined service behavior:

    cl-2-1) Illegal transitions when services are combined.

    cl-2-2) Lost transitions when services are combined.

    cl-2-3) Illegal states when services are combined.

    cl-2-4) Lost states when services are combined.

cl-2-5) Duplicated definition of terminology by multiple designers.

In the service specification design stage, the properties mentioned above can appear as service interactions such as deadlock, looping, non-determinacy of transition and transition to an abnormal state. Therefore, a concrete classification in the specification design stage can be regarded as deadlock, loop, non-determinacy, transition to an abnormal state and duplicated terminology. The classification discussed here is shown in table 1.

Table 1 Classification of service interactions

Classification	FSM viewpoints	Examples
Deadlock	cl-1-1, cl-2-2, cl-2-4	(1),
Loop	cl-1-1	(1)
Non-determinacy	cl-1-2	(2), (4), (8)
Transition to an abnormal state	cl-2-1, cl-2-3	(3), (5), (6), (7), (9)
Duplicated terminology	cl-2-5	(10)

cl-i-j: defined in section 3.1   (k): described in section 3.4

3.2 Service Model to be Discussed

The service model treated in this paper, is described as follows:

(1) The service specification is treated as a set of behaviors that are independent of network behavior, which is treated as a black box (Fig. 1),

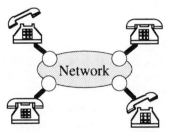

Network is assumed to be a black box.

Fig.1 Service model

(2) and as a set of behaviors among multiple terminals (Fig. 2).

Figure 2 indicates the behavior between terminals P and Q, where terminal P is in the hearing dial-tone state ("dial-tone(P)") and terminal Q is in the idle state ("idle(Q)"). If terminal P dials terminal Q, the next state becomes a calling state from terminal P to terminal Q, where terminal P is in the hearing ringback-tone state to terminal Q ("ringback(P, Q)"), and terminal Q is in the hearing ringing-tone state from terminal P ("ringing(Q, P)").

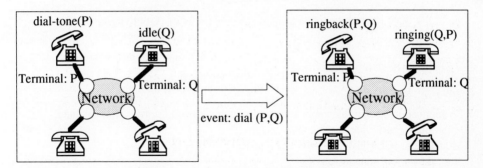

Fig.2 Transition among multiple terminals

3.3 STR

As a service specification description language, ATR developed a new language called "STR(State Transition Rule)" [6] which is based on state transition rules. A feature of the STR is that a service behavior that does not depend on the target system's behavior is described from the user's viewpoint. A service specification is constructed as a set of features that are described using two types of forms: rule description and inhibited primitive set description. First, rule description and inhibited primitive set description are explained; second, the applications of the rule are explained.

3.3.1 Rule Description and Inhibited Primitive Set Description

The definition of rule description is given as follows:
[Definition 1] (rule description)
  $R_i = IS_i + E_i + NS_i$
  $IS_i, NS_i = \{SP_1,...,SP_n\}$
    R_i: a rule that consists of IS_i, E_i, and NS_i,
    IS_i: the current state of rule R_i,
    E_i: the event of rule R_i,
    NS_i: the next state of rule R_i,
    SP_i: the state description primitive, which consists of the identification
        of the state and the terminal. For example, $SP_i(A, B)$ indicates that
        terminal A is in the SP_i state with terminal B. "A" and "B" are
        treated as terminal variables.

  A rule is described as a tuple of the current state, the event, and the next state. The current state and the event are treated as a condition of rule application. The current state and the next state are described as a set of state description primitives. In a rule, different terminal variables indicate different terminals.

  The definition of inhibited primitive set description is as follows:
[Definition 2] (inhibited primitive set description)
  {RS} (RN)
  RS, RN= $\{SP_1,...,SP_n\}$
    RS: a combination of state description primitives that is inhibited in
        the actual state, is a condition when an inhibited primitive set is
        applied.

RN: the state for making the next state when an inhibited primitive set is applied; if RN is not described, the current actual state does not change.
SPi: the state description primitive (Def. 1).

3.3.2 Rule Application
Rule applications: basic rule application and rule application priority, are explained.
[Definition 3] (basic rule application)
$(ISi \subseteq G) \wedge (GE = Ei) \rightarrow Ri$
ISi: the current state of rule Ri,
G: the actual state,
GE: an event that occurs in state G,
Ei: an event of rule Ri,
Ri: a rule that is applied.
[Definition 4] (rule application priority)
$\{(ISi \tilde{O} G) \wedge (GE = Ei)\} \wedge \{(ISj \subseteq G) \wedge (GE = Ej)\} \wedge (ISj \subset ISi) \rightarrow Ri$
ISi, G, GE, and Ei are the same as those in Def. 3,
ISj: the current state of rule Rj,
Ej: the event of rule Rj.
If "$(ISi \subseteq G) \wedge (GE = Ei)$", "$(ISj \subseteq G) \wedge (GE = Ej)$" and ("$ISj \subset ISi$") are satisfied, then when rules Ri and Rj are applicable, the former rule (Ri) is applied prior to the latter.

The following are examples of rule description.
pots-1) dial-tone(A), idle(B) dial(A, B): ringback(A, B), ringing(B, A).
cfv-1) dial-tone(A), idle(B), m-cfv(B, C), idle(C) dial(A, B):
         pingring(B, A), m-cfv(B, C), ringback(A, C), ringing(C, A).

The above "pots-1)" means that the rule is the first of the features of POTS (Plain Old Telephone Service) and "cfv-1)" means that the rule is the first of the features of CFV (Call Forwarding Variables). The above "dial-tone(A)," "idle(B)," "ringback(A, B)," "ringing(B, A)," "m-cfv(B, C)," and "pingring(B, A)" are called state description primitives. "dial(A, B)" is called the event description.

For example, the behavior shown in Fig. 2 is described as "pots-1)" above, which means that when terminal A is in the dial-tone state and terminal B is in the idle state, if terminal A dials terminal B ("dial(A, B)"), the next state becomes a calling state from terminal A to terminal B, where terminal A is in the hearing ringback-tone state to terminal B ("ringback(A, B)"), and terminal B is in the hearing ringing-tone state from terminal A

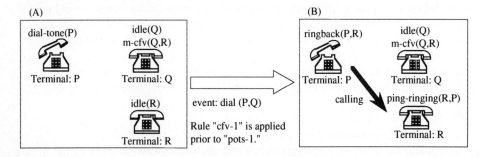

Fig.3 Example of rule application

("ringing(B, A)"). Here, "pots-1" is applied by identifying A as being P, and B as being Q (Fig. 2). If "cfv-1" is applicable, "pots-1" is also applicable because the current state of "pots-1" is a subset of the current state of "cfv-1". In this case, "cfv-1" is applied prior to "pots-1" according to the rule application priority (Def. 4). The transition by the application of rule "cfv-1" above is shown in Fig. 3. In Fig. 3-(A), if terminal P dials terminal Q, the current state and event of "cfv-1" are judged to be satisfied by identifying A as being P, B as being Q, and C as being R.

3.4 Detailed Explanations of Service Interactions

Detailed explanations including some examples of the service interactions described in section 3.1 and in table 1 are given in this section.

3.4.1 Logical Problems of the FSM
(1) *A state that has no succeeding transitions*:
If one of the actual states is a state that has no succeeding transitions, it is impossible to leave that state, and kind of deadlock occurs at the service specification level.

If there are states that never return to the initial state, a transition loop occurs among these states. For example, in Fig. 4, states "S4," "S5," and "S6" are states that never return to the initial state "S0."

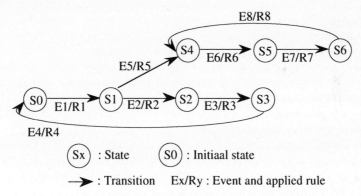

Fig.4 Example of transitions

(2) *A state that has multiple transitions for one event*
In some cases, as a result of the synthesis of services, a state has more than one transition for the same event. Suppose that a terminal activates call waiting service and call forwarding service simultaneously. In this case, while the terminal is in the talking state with another terminal, a new call is terminated to the terminal. Should the terminal invoke call waiting service and transit to the cw-ring state, or should it invoke to call forwarding service and remain in the same state? This is one example of transition non-determinacy.

3.4.2 Semantical Problems of the FSM
(3) *Prevention of transition*:
Although a service feature is applicable, transition is prevented by some other service feature.

For example, assume the combination of TWC (Three-Way Calling) and emergency service. With TWC, a terminal enters the three-way calling mode after putting the current conversation terminal on hold, whereas, with emergency service, a terminal rejects being put on hold. Assume terminal P has three-way calling capability and is connected with terminal Q which has emergency service capability. In this case, terminal P cannot be in the three-way calling mode (Fig. 5). This is classified as a transition to an abnormal state.

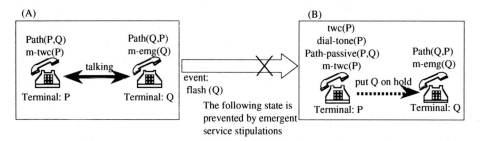

Fig.5 Typical prevention of transition

(4) *Intentional conflict in rule application*
If a service specification is applied prior to some applicable service specification, the designer must judge whether the transition is correct from the standpoint of the combined service specifications because the intentions of the different services conflict. This is classified as non-determinacy.
(5) *Lost transitions of the new service and the existing services*:
Transitions of the new service may be excluded by the existing service specifications, or transitions of the existing service may be excluded by the new service specification. These excluded transitions should be analyzed because transitions of the new service and the existing services are guaranteed even if the new and existing services are combined. This is classified as transition to an abnormal state.

Suppose that the current state of a rule(Re) in the existing service is ISe and that of a rule(Rn) in a new service is ISn. If ("ISe Ã ISn") is satisfied, Re will never be applied. Therefore, in this case, the corresponding transition of the existing service is lost.
(6) *Logically illegal state*:
A logically illegal state is a state that must not appear in the actual state. i.e., 1) a duplication of the same state as a terminal state, 2) a combination of the exclusive states as a terminal state, and 3) an illegal state among multiple terminal states. For example, the combination of the idle state and the dial-tone state as a terminal state is illegal because it should not be the state of a terminal. If logically illegal states appear in the generated transitions, they must be detected. This is classified as transition to an abnormal state.
(7) *Undefined state*:
An undefined state is a state that never appears in each individual service. Suppose, when a new service is added to an existing service, nondeterminacy occurs. In this situation, when a rule(Rn) of the new service is applied, the next state of the exsisting service consists of two parts, one is part of the next state of the new service and the other is part of the current state of the existing service. If the next state of the existing service can not be found in the exisiting service states that appear before a new service is added, this state is called an undefined state. Figure 6 shows this situation, which is classified as a transition to an abnormal state.

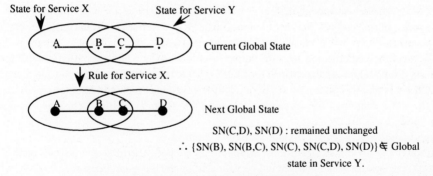

Fig. 6 Transition to Undefined State

(8) Vagueness in additional and undefined transitions
Vagueness of specification will occur when services are combined, since each service specification is designed without considering, or with poor consideration of, the other services.
For example, consider the combination of CCBS (Completion of Call for Busy Subscriber)

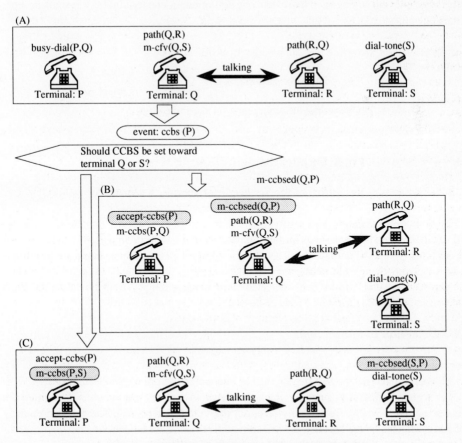

Fig.7 Vagueness between CFV and CCBS

and CFV (Call Forwarding Variable).

Assume terminal Q has preset CFV to terminal R, and terminal R is in the busy state. In this situation, if terminal P dials terminal Q, terminal P enters the busy state, since the dialed call is forwarded and the call forwarding terminal R is in the busy state. If terminal P then requests CCBS, although the request for CCBS is set to terminal Q according to the features of CCBS, whether the request for CCBS is set to terminal Q or R should be determined when designing the combined services of CCBS and CFV (Fig. 7). This is classified as non-determinacy.

(9) *Additional and undefined transitions to be validated*:
Additional and undefined transitions, which are generated when the new service and the existing services are combined, should be validated by the designer, because
these transitions are not defined in the individual service transitions. Thus, these transitions as well as the individual service transitions need to be checked by the designer. The unwanted states and transitions may be included in additional and undefined transitions. This is classified as transition to an abnormal state.

(10) *Duplicated definitions of terminology*
When multiple designers participate in the development or the designer of a new service is different from that of existing service, duplicated definitions of terminology can occur. In this case, one object may have more than one definition, or one terminology designates more than one object (Fig. 8).

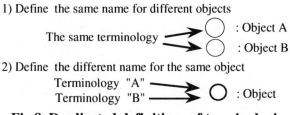

Fig.8 Duplicated definitions of terminologies

4. Detection and Resolution of the Service Interactions

In this section, detecting and resolving support methods for service interactions are described. The last example (10) in the previous section is a very important issue, one that causes many kinds of interactions. Therefore, it is important to eliminate duplicated terminology before checking each interaction.

As deadlock and loop are easily detected, only, nondeterminacy, transition to an abnormal state and duplicated terminology are discussed here.

4.1 Detection of Interactions

(1) Non-determinacy
Non-determinacy means that the applicable rule cannot be uniquely selected. In specifications written using STR, rules in which the provided services are individually written are collected into a whole, from which the applicable rule is determined. The non-determinacy error is detected by the fact that the selection of this rule is not unique.

Accordingly, the detection algorithm proceeds as follows.

1) A pair of arbitrary rules that have the same event is detected.
2) If the conjunction of the current states of the rules is not included in the actual global state, this pair does not cause non-determinacy.
3) If the conjunction of the current states of the rules is included in the actual global state, compare the current state of both rules.

If either state includes the other state, no non-determinacy occurs. If neither state includes the other state, non-determinacy is detected.

(2) Transition to an abnormal state

A transition to a logically illegal state is easily detected by checking the state whenever a rule is applied, whether the state is logically illegal or not. Therefore, here, transition to an undefined state is described. In order to simplify our explanation, we describe only a case in which two services are offered simultaneously. An abnormal state transition occurs due to service interaction because there is no relation between the current states of the services' two rules (global state descriptions). When the two rules compete, either may take priority and the next state of the service with the non-applied rule will not be reached; thus, a transition to an undefined state occurs.

A global state consists of state primitive S(I), which denotes the state of terminal I, and state primitive S(I,J), which denotes the state of the relation between terminals I and J.

Accordingly, the detection algorithm proceeds as follows.
1) Rules with current states in which no inclusive relationship exists with the same event, are detected.
2) Whichever rule applies, the next actual state of the service to which the rule applies is compared with that of the single service.
3) If the comparison result in 2) does not yield coinciding states, a transition to an undefined state is detected.

(3) Duplicated terminology

To detect duplicated terminology, it is necessary to know the meaning of the words. The

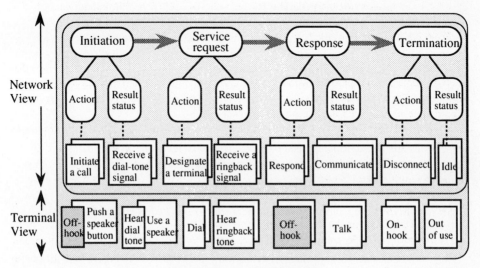

Fig.9 Conceptual Model

meaning of the words depends on the situation in which the words are used. We developed a conceptual model for switching system functionality(Fig. 9). The definitions in this model do not depend on the system or designer.

Therefore, in our proposed system, the situation words in which used by a designer are confirmed by asking a designer. If the situation is fixed, then, the words are compared with words related to the situation. If the word is confirmed to have the same meaning as the word in the model, the duplicate definition is detected.

4.2.3 Resolution of interactions
(1) Non-determinacy
As shown in the example of the error of non-determinacy in (1) in 4.2.2, in most cases, the rules that take priority are the service specifications and, according to the user's instructions, a rule with priority will be generated. The algorithm generates a rule that replaces the current state of the priority rule with the disjunction of the current states of both competing rules(Fig. 10).

(1) CW takes priority
dial-tone(C), path(A,B), m-cw(A), idle(D), m-cfv(A,D)
dial(C,A) : ringback(C,B), ringing(B,C), path(A,B), m-cw(A), idle(D), m-cfv(A,D)

(2) CFV takes priority
dial-tone(C), path(A,D), m-cw(A), idle(D), m-cfv(A,D)
dial(C,A) : ringback(C,D), ringing(D,C), path(A,B), m-cw(A), m-cfv(A,D)

Fig. 10 Resolution of non-determinacy

(2) Transition to undefined state
Because it is important to guarantee correct state transitions, we add the following rules to the existing ones (taking the case in which only a single service is offered for consideration, the existing rules remain unchanged). The current state of each new rule is equivalent to the disjunction of the current states of the competing two rules.

There are five state primitives (called primitives hereafter) that are related to service Y in Figure 6. S(D) is a primitive of terminal D, which emerges only in service Y. S(C) is a primitive of terminal C that also emerges in service X in Figure 6 and has a relation with terminal D. S(C,D) is a primitive of the relation between terminals C and D. S(B) is a primitive of terminal B, which also emerges in service X but has no relation with terminal D. S(B,C) is a primitive of the relation between terminals B and C.

As shown in Figure 6, S(D) and S(C,D) remain unchanged. And although S(C) is changed to the next state, considering the relation with terminal D, it may be an incorrect state. Therefore, the next state of a rule, which resolves the transition to the undefined state, is defined as follows.
{S(A),S(A,B),S(B),S(B,C),S'(C),S'(C,D),S'(D)}

S(I) and S(I,J) are equal to those in the next state (Figure 6). S'(I) and S'(I,J) satisfy the following conditions.
1) {S(B),S(B,C),S'(C),S'(C,D),S'(D)} is included in the actual global state of service Y.
2) There is at least one rule that changes the current state of service Y (Figure 6) to the global state defined in 1).

The next state of the rule, which resolves the abnormal state transition, is shown in Fig. 11.

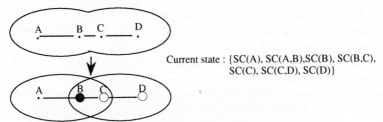

Current state : {SC(A), SC(A,B),SC(B), SC(B,C), SC(C), SC(C,D), SC(D)}

Next State : {SN(A), SN(A,B), SN(B), SN(B,C), S'N(C), S'N(C,D),S'N(D)}
where SN(A), SN(A,B), SN(B), and SN(B,C) are equal to
those in Fig.6.
S'N(C), S'N(C,D), S'N(D), satisfy the following.
{SN(B), SN(B,C), S'N(C),S'N(C,D), S'N(D)} ∈ Global State of
Service Y

Fig. 11 Resolution for Transition to Undefined State

(3) Duplicated definitions of terminology
As to duplicated definitions of terminology, just after detection mentioned in the previous section, the word used by the designer is rewritten using the corresponding word in the conceptual model.

5. Some Considerations on our Experiment

The point (state) where service interaction occurs is not always the point that needs to be resolved, since service interactions should be resolved from the standpoint of the combined services. This is almost equivalent to debugging in programming. Thus, grasping the execution sequences from the initial state to the inconsistent point and prompting the designer to analyze the actual reasons are useful. By using the visual programming method [7], which was also developed at ATR, as an interpreter of service descriptions described with the STR method, tracing the execution sequences from the initial state to the inconsistent point is possible. Using the execution sequences a rapid prototyping is proposed as a way to validate and verify the service specifications [8]. Thus, after grasping the execution sequences, by using the visual programming method mentioned above, the designer can visually analyze the actual cause of the service interactions.

6. Conclusion

A general framework for classifying the telecommunication service interactions and classification of the service interactions in the service specification design stage is proposed. Some detailed examples of the interactions are also given. Support methods for detecting and resolving the interactions have been proposed. These methods have been implemented, and are now under evaluation. Twenty-eight of the services in Bellcore LSSGR and CCITT service specifications, including UPT, are now being tested.

Acknowledgment
The authors wish to thank Dr. Habara and Dr. Terashima for their encouragement. They also wish to thank their colleagues for helpful discussions.

References

[1] T. F. Bowen, et al., "The Feature Interaction Problem in Telecommunications Systems," Proc. of the 7th IEE Int. Conf. on Soft. Eng. for Telecom. Systems, pp. 59-62, July, 1989

[2] Jane Cameron, Nancy Griffeth et al., "A Feature-Interaction Benchmark for IN and Beyond". IEEE Communication Magazine, March 1993

[3] Y. Wakahara, et al., "A Model for Detecting Interactions Among Services and Service Features," Proc. of Int. Workshop on Feature Interactions in Telecommunications Software Systems, pp. 82-94, 1992

[4] E. Kuisch, et al., "A Practical Approach Towards Service Interactions," Proc. of Int. Workshop on Feature Interactions in Telecommunications Software Systems, pp. 41-59, 1992

[5] Y. Harada, K. Takami, T. Ohta, and N. Terashima, "A Conflict Detection Method for Telecommunication Billing Specification Descriptions," IJCA Tran. Vol. 34, No. 5, pp. 1064-1073, 1993

[6] Y. Hirakawa, Y. Harada, and T. Takenaka, "Behavior Description for a System which Consists of an Infinite Number of Processes," BILKENT Int. Conf. on New Trends in Communication, Control, and Signal Proc., pp. 59-68, 1990

[7] K. Takami, Y. Harada, T. Ohta, and N. Terashima, "A Visual Design Support System for Telecommunication Services," Proc. of IPCCC, pp. 593-599, 1993

[8] T. Ohta, K. Takami, A. Takura, "Acquisition of Service Specifications in Two Stages and Detection/Resolution of Feature Interactions," Proc. of TINA'93, vol.2 pp.173-187, 1993

Feature Interactions among Pan-European Services

Kristofer KIMBLER
TTS, Lund Institute of Technology, Box 118, 221 00 Lund, Sweden,

Eric KUISCH
PTT Research, P.O.Box 421, 2260 AK Leidschendam, The Netherlands

Jacques MULLER
FRANCE-TELECOM - CNET, 92131 Issy-les-Moulineaux, France

Abstract. This paper presents results of the research on feature interaction performed in PEIN, a EURESCOM research project. The objective of the paper is to communicate the practical experience of the project with feature interactions and to discuss the technical approach used by the project.

1. Introduction

This paper presents results on feature interactions obtained by the PEIN project. The PEIN project consortium consists of 14 public network operators [1] (PNOs) and is supervised by Eurescom, the European research institute for strategic studies in telecommunications.

The project aims to enable **Pan-European** services by co-operation between **I**ntelligent **N**etwork platforms of public network operators in Europe (hence the acronym PEIN). The target is that these services can be provided throughout Europe within the 1996-1998 time frame after the project has completed its research activities in 1995. The PEIN project has identified five services which have a big market potential when introduced as Pan-European services (so-called PEIN services). They are: Freephone (FPH), Premium Rate (PRM), Charge Card Calling (CCC), Virtual Card Calling (VCC) and Virtual Private Network (VPN).

Introduction of these services on a Pan-European level is a formidable task which involves a lot of different technical problems. Among the issues which have to be tackled are: specification of architectural requirements to support the co-operation among the IN platforms, detailed specification of the identified services and the usability of the services

---

[1] ATC Finland, BT, SIP/CSELT, Deutsche Bundespost Telekom, France-Telecom, Norwegian Telecom, Royal PTT Netherlands, Telecom Eireann, Telecom Finland, Telecom Portugal, Tele Denmark, Telefones de Lisboa e Porto, Telefónica de España, Telia.

from user's point of view. In this paper we focus on one other major issue: feature interactions related to provisioning of the Pan-European services.

1.1. PEIN Research on Feature Interaction

The problem of unpredicted and undesirable feature interactions is a threat to the provisioning of Pan-European IN services. The project identified a need for solutions, methods and tools helping to handle different aspects of this problem, especially as the PEIN services are processed in different networks. The objective of PEIN with respect to research on feature interaction is to:

- Analyse feature interaction related to the pan-European services identified in the project,
- Define guidelines how to handle feature interaction related to these services,
- Define general guidelines on feature interaction occurrence and avoidance with respect to introduction of new services, and
- Develop a tool-supported method for spotting of feature interaction in order to enable that the PEIN services can be introduced by PNOs on their IN platforms in an interaction-free way.

The work is still on-going; in this paper the results on the first objective are presented: an analysis method for feature interaction. Even if only five services need to be analysed, the great number of service features (the PEIN list of features contains over forty items!) make the analysis of interactions a highly non-trivial activity. Especially as PEIN considered a diverse set of service features like: Origin Dependent Routing, Queuing (Freephone features), Pre-Defined Destination, Language Selection (Card Calling features) and Outgoing Call Screening (Virtual Private Network feature). The project therefore needed to identify a pragmatic and well-structured working method for analysing and spotting feature interaction related to the PEIN services in order to produce a consistent, manageable and non-ambiguous survey of feature interaction problems.

An iterative approach for the feature interaction spotting was chosen: the project agreed on analysis principles, then these principles were applied to analyse feature interaction related to the PEIN services leading to enhancement of the analysis principles which could then be applied again for a more detailed feature interaction analysis. This process is pursued until the analysis principles are considered mature enough and need no major enhancements. In the sequel the PEIN principles for feature interaction analysis are presented as well as an outline of the analysis method itself.

2. Objectives of the PEIN Analysis Method

2.1. Motivation for Systematic Analysis of Feature Interactions

There are many claims that the analysis of feature interactions is almost impossible for one simple reason: the number of interaction cases, or rather combinations, to be analysed grows exponentially with the number of the features [1]. From the purely combinatorial point of view this statement is true, but it might not be necessary to analyse all the combinations. When all combinations of features that can never lead to any interaction have been eliminated, only a small factor of all the cases is left for further analysis and resolution. The main idea of the PEIN approach was: discard irrelevant combinations of features and analyse the remaining ones.

To meet this objective a method had to be created for distinguishing interaction-prone feature combinations from those which can never cause interactions. The second issue was to find a technique for analysing these interaction-prone feature combinations and spotting possible interactions.

2.2. Spotting of Interaction-prone Feature Combinations

The approach for finding interaction-prone combinations of features is based on two simple observations.
- Many features show similar characteristics.
- Features are not stand-alone entities.

We can group features into categories like restriction, routing, charging, etc. Some of these categories have, by their nature, nothing in common with one another, which means that any possible combination of their members will be interaction-free. For instance, restriction features (e.g. call screening) never interact with user dialogue features (e.g. outgoing user prompter). The analysis of possible interactions between the feature categories scales down the complexity of the problem dramatically, because we deal with a very limited number of combinations here.

Features are packaged into services. Therefore, possible feature interactions should always be considered and analysed in a service context. One could expect that this approach will make the task even more complex by increasing the number of interaction cases to be analysed, but the result is just the opposite. If two services don't interact within the same call, no interaction between their features can occur.

2.2.1. Interactions among Feature Categories

The classification of the features into the categories is based on the similarities of their functionality, and the similarities of the roles they play towards the user and towards the other features. More discussion about the feature categories can be found in section 3.2.3.

The decision as to which combinations of the categories are interaction-prone and which are not is done upon the roles they play and resources they use. If one category plays some role towards another one (e.g. restriction features constrain numbering features) or the two categories access or modify the same resources (e.g. restriction and access features use calling line identification data) this means that their members might interact. According to this criterion, the combination of a category with itself should be always regarded as interaction-prone because the features belonging to the category have similar characteristics and they use the same resources. Case study can be used as a valuable supplementary technique here. If at least one interaction case involving members of two different categories can be found using heuristic methods, the combination of the categories should be regarded as interaction-prone.

2.2.2. Feature Combinations in Service Context

The identification of service contexts for each interaction-prone combination of the features is based on use-case driven analysis. This kind of technique aims at analysing the system, in our case Pan-European IN, starting from the user's point of view, i.e. finding different possible ways how IN services could be used and combined by potential service subscribers and users. This kind of requirements analysis technique is widely used in Object Oriented Software Engineering [6,7].

During the analysis actors and use-cases are defined and described. An actor models a role played by a system user. We can say that each actor models different needs and different behaviour of the potential system users. Actors can model also roles of non-human users, i.e. other systems or devices interacting with the system in question.

By analysing the potential behaviour of different actors (market analysis statistics are very useful for that purpose) we are able to identify several use-cases. A use case is one specific scenario of the system usage described as a sequence of events and user interactions with the system (e.g. by means of Message Sequence Charts). For the purpose of use-case driven analysis, an IN service is to be treated a function offered to the user as a whole, rather then as a loose collections of separate network capabilities (features).

The analysis results in a service usage model consisting of different use-cases. This model is primarily used to determine interactions on the service level, or in other words, to determine which service combinations are possible within the same call. This kind of information is then utilised for finding a service context for each pair of features. A given combination of features (F1,F2) can cause an interaction if and only if there exist two services S1 and S2 such that F1 belongs to S1 and F2 belongs to S2, and both S1 and S2 can be used within the same call. If such a pair of services (S1,S2) exists, it is called a service context for the combination of features (F1,F2).

The service usage model can be utilised by the service provider as feedback to the market analysis. For instance, some of the discovered service combinations like using VPN member lines for FPH termination lines could be very attractive for certain groups of business subscribers (e.g. airlines would like to use VPN for internal communication and FPH for place booking service). The usage model can be also used by the service provider as an input to the policy-making process. Some odd combinations of the services which might cause serious conceptual and technical problems could be banned this way (using charge card for calling free-phone number or initiating virtual charge call from within a charge card service).

2.2.3. Spotting Algorithm

The algorithm for spotting interaction-prone combinations of the features requires information about the interactions among the feature categories and the interactions among the services. The algorithm aims at sifting out interaction-prone combinations of the features. The algorithm could be described in a quasi-formal way like this:

```
FOR all pairs of features (F1,F2) DO
    IF interact (category (F1), category (F2)) THEN
        FOR all pairs of services (S1,S2) DO
            IF member (F1,S1) & member (F2,S2) & interact (S1, S2) THEN
                analyse (F1,F2) in context of (S1,S2)
```

The combination of the features which pass through the sieve have to be analysed. The analysis applied for this purpose is described in section 3.2.4. The algorithm presented above, being rather simple, could be incorporated into a tool, for instance an expert system, supporting the analysis process; such work is actually planned within the PEIN project.

3. Outline of the PEIN Analysis Method

3.1 Role of Service Life Cycle in the Analysis

The PEIN project considered it useful to use a common framework for describing a services. To this end, the service life cycle was adopted [2] which is currently developed by ETSI/NA6 [2]. We refer to this document for more details on the service life cycle concepts and terminology, but most of the terminology is well-known or self-explanatory. The service life cycle stipulates that a service, during its life, goes through some stable states, and that these states are separated by steps (or transitions) which permit passing from one state to another. The states and transitions will be presented in more detail in the sequel.

Figure 1 shows the pertinent area for P230 interaction studies.

Some of the examples met in the technical literature show that two services may present several different interaction cases (see [8]), depending either on the event occurring in the processing of the first service and on the considered phase of the second service. Hence we have to consider all the combinations between the event of a service S1 (i.e. a transition earlier described) and the state of a service S2. An exhaustive interaction study will consist of investigating the impacts of five S1 transitions (initialisation, activation, invocation, end of execution, deactivation) and three stationary S1 transitions (modification of initialisation data, modification of activation data and modification of invocation data) on four S2 states: subscribed, initialised, activated and invoked. This is shown in figure 2.

---

[2] ETSI is the European Telecommunications Standards Institute, NA6 is the Sub-Technical Committee within ETSI responsible for standardisation of the Intelligent Network

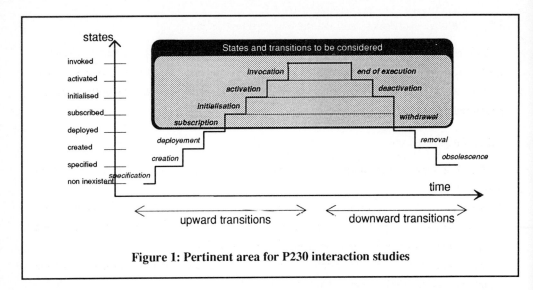

Figure 1: Pertinent area for P230 interaction studies

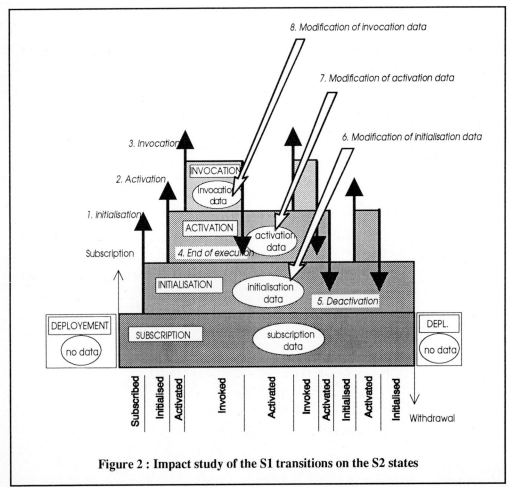

Figure 2 : Impact study of the S1 transitions on the S2 states

3.2. PEIN's Approach towards Feature Interaction Analysis

The feature interaction study in PEIN went through the following stages: simplification of the service life cycle, introduction of service topologies (configurations), selection and classification of features to be studied and finally analysis of the services and features using a structured method. Each of the stages will be described below.

3.2.1. Simplifications

The following simplifications to the above mentioned state/transition model have been applied:
- not to consider the end of execution nor the modification of invocation data, supposed not to bring interaction problems (this is an axiom, or a simplifying choice, and it hasn't been demonstrated as true, but no counter-example has been found);
- merge initialisation, modification of initialisation data and modification of activation data into a single transition called parameterisation;
- not to consider the subscribed state;
- not to consider the deactivation transition.

The table showing the remaining states and transitions is shown hereafter. It is to be filled when considering the interaction of a feature F1 with a feature F2.

F1.F2	F2.activation	F2.parameterisation	F2.invocation
F1.initialised			
F1.activated			
F1.invoked			

3.2.2. Call Configurations

Three configurations of calls have been introduced, in order to allow a better typology of the interaction cases.

Configuration 1: Two destinations are involved in the call, services S1 and S2 are allocated to the same destination: calling destination (A) or called destination (B).

Configuration 2: Two destinations are involved in the call, service S1 is allocated to the calling destination (A) and service S2 is allocated to the called destination (B).

Configuration 3: Three destinations are involved in the call (A,B,C):
S1 is allocated to A, S2 is allocated to C; or,
S1 is allocated to B, S2 is allocated to C; or,
S1 is allocated to A, S2 is allocated to B

An example of configuration 3 is the following: S1 is Call Forwarding Unconditional, allocated to B, and S2 is Automatic Call-Back, allocated to A. If B calls A who uses ACB, C will receive a call. Refer to [3] or [8] for a complete study of this interaction case.

3.2.3. Definition and categorisation of service features

The features taken into consideration have been defined either by ETSI/NA1 [3] or by the EURESCOM project P101 [5]. They are documented in [4].

A classification of features in feature categories was considered useful as general considerations can be used to determine the interaction among different categories. Ideally, logical analysis of the characteristics of each feature category should lead to dismissal of certain category combinations where interaction is unlikely to occur. This proved quite difficult in practice and the project has investigated the combinations of feature categories both with logical arguments and by analysis of specific feature combinations.

The classification of feature categories in [4] is as follows:

- charging features,
- restriction features,
- routing features,
- customisation features,
- information management features,
- numbering features,
- access features,
- user dialogue features.

3.2.4 Structured Analysis Method

For this structured approach, the idea has been retained that feature interactions should be always considered and analysed in a service context. Considering that if two services do not interact, i.e. cannot be invoked, parameterised, activated, etc. within the same call, no interaction between their features can occur either, it has been decided to look for a method allowing to eliminate as far as possible these "harmless" combinations. The PEIN analysis method consists of the following actions:

Action 1	Analyse interaction between service pairs
Action 2	Analyse combinations of feature categories
Action 3	Compare stand-alone feature pairs
Action 4	Compare feature pairs within service context

All these actions, as well as their relationships, are shown in figure 3.

---

[3] ETSI is the European Telecommunications Standards Institute, NA1 is the Sub-Technical Committee within ETSI responsible for standardisation of IN, ISDN and B-ISDN services.

Action 1: Analyse interaction between service pairs

The services in their different configurations, and only with their core features, are compared with each other in order to determine interaction (either positive or negative) on the service level.

Result: A table indicating the configuration in which the interaction among services occurs, if any (so-called *Service Interaction Table* in the following).

Action 2: Analyse combinations of feature categories

For each combination of feature categories, it is determined whether there is interaction between feature categories. The reason why two groups are considered as not interacting are to be given with logical reasoning, and not after a heuristic consideration of the features included in both groups.

Result: A table classifying the chance that a certain combination of two groups causes interaction.

Action 3: Compare stand-alone feature pairs

The stand-alone interactions between features belonging to interacting groups are analysed. For every combination of groups with a high probability level (determined by action 2), the pertaining features are investigated as stand-alone entities, i.e. in a service-independent context.

Result: A table stating whether or not the considered feature may interact.

Action 4: Compare feature pairs within service context

For every feature combination which causes interaction according to action 3, a detailed investigation is performed:

1. All PEIN services which contain feature F1 and F2 respectively are selected.

2. The *Service Interaction Table* is used as a mask to eliminate those service combinations which are irrelevant because of lack of interaction on the service level.

3. The configurations in the *Service Interaction Table*, which are not applicable to the specific pair of features are discarded (e.g. if F1 and F2 are both terminating features then configuration 2 is discarded).

4. Perform full analysis of interactions for the remaining configurations using a state/transition table. Collect results in the *Service Feature Interaction Table* (results are described in prose in annotations to the table) using a standard coding and the following format:

F1 - F2	S1	S2	S3	S4	S5
S1					
S2					
S3				**code**	
S4					
S5					

The codes used for describing analysis results has the following format:

```
code              ::=    {state}{transition}{configuration},
{state}           ::=    [ a - activated | p - initialised | i - invoked ]
{transition}      ::=    [ a - activation | p - parameterisation | i - invocation ]
{configuration}   ::=    [ 1 | 2 | 3 ]
```

The *Service Feature Interaction Tables* are not symmetrical. So for each F1-F2 combination, there are potentially two *Service Feature Interaction Tables*, F1-F2 and F2-F1. If the order of the features is significant for certain combinations, the codes for reverse configurations are to be introduced into the existing *Service Feature Interaction Table* in the following way:

ii2   - F1.invoked/F2.invokation   (lower-case code)
IP2   - F2.invoked/F1.parametrisation   (upper-case code)

The application of two features to the same service is to be separated from the application of these two features to two different services, for instance by a specific indication in the tables. Such indication must help to differentiate three cases:

Fx/Si vs. Fx/Sj
Fx/Si vs. Fx/Si on the same call   (indication: = s)
Fx/Si vs. Fx/Si on two different calls   (indication: = d)

Interaction between two features is analysed in four steps:

1. F1 applied to Sn in a given state is considered with F2 applied to Sm at a given transition,
2. F2 applied to Sn in a given state is considered with F1 applied to Sm at a given transition,
3. F1 applied to Sm in a given state is considered with F2 applied to Sn at a given transition,
4. F2 applied to Sm in a given state is considered with F1 applied to Sn at a given transition.

Results of the first two steps are described together in one entry (and marked by the lower/upper case convention), while the last two steps are described in another entry (using the same convention).

Summarising all actions in the PEIN analysis method, as well as their relationships, we arrive at figure 3.

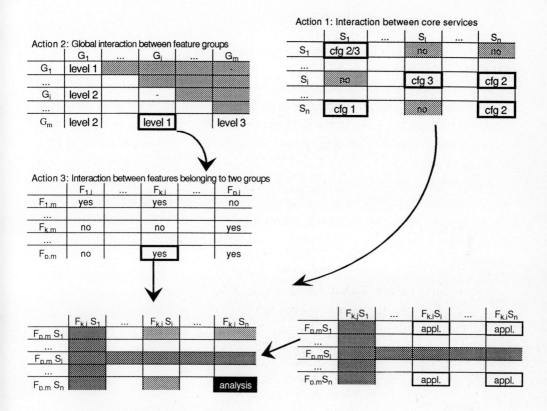

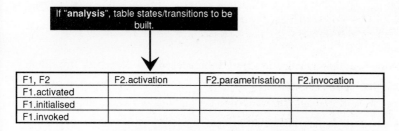

Figure 3: Global view on the PEIN analysis process

3.3. Example of Applying the PEIN Analysis Method

Consider two features, Call Queuing (QUE) and Alternative Destination Routing on busy (ADR).

Both are routing features. Action 2 shows that there is interaction between the routing feature category and itself, so the feature analysis process will get to Action 4 and has to determine the applicable services and configurations in which analysis has to be performed. The QUE feature applies to the PEIN services FPH and PRM, while ADR applies to FPH, PRM and VPN.

From Action 1 the configurations are derived which have to be considered for every service combination of PEIN services. For example, Action 1 shows that we have to consider the QUE-ADR interaction for the service combination FPH-VPN in configuration 1. Configuration 1 in this context means that we have to consider a FPH call to a VPN destination.

Once these results are available, Action 4 can be performed and the state/transition table is built.

One of the results is that we find that the invocation transition of VPN, due to the ADR feature, affects the invoked state of FPH as the ADR feature reroutes the call and bypasses the QUE feature (i.e. ADR disables the modification of FPH invoked state by the QUE feature).

4. Conclusions

- The experience within the PEIN project shows that an iterative approach can be useful to derive a pragmatic but also well-structured method for feature interaction analysis. The results of applying such a method by PEIN were satisfactory and many feature interaction cases could be spotted.
- The work of PEIN on feature interaction needs to be broadened and deepened. Breadth is needed in the sense that interaction among the PEIN services and already existing switch-based services needs to be addressed. This will most certainly have a major impact on the current analysis method for feature interaction. Depth is needed as there have only been two analysis phases until now and major improvements are expected.
- Interaction spotting is highly dependent on sufficiently detailed service and feature descriptions. In fact the feature interaction problem is very much a feature description problem and puts high requirements on specification methods. Special attention should be given to specification methods for composing services out of service features. Lack of decent methods will give rise to serious feature interaction problems among features within a single service.
- Feature interaction is a highly complex problem and influences several quite different parts of PNO's organisations e.g. marketing, service specification, service testing and service implementation. Methods developed to handle feature interaction are so

complicated that they can only lead to practical application by this broad range of people in PNO's organisations if they are supported by software tools such as tutorials, expert systems and simulation programs.
- The treatment of an interaction case has to be considered in two different steps [8]:
  - as a first step, during the specification phase, the service provider's technical staff analyses the interworking between the two features and deduces the interaction cases which may be a problem for any of the actors;
  - as a second step, both technical and commercial staffs make choices from the possible solutions, according to network capabilities, human factors constraints, sociological data, economic factors and users' requirements.

The P230 interaction spotting method is only intended to perform the first step. Engineers working in the project are not committed to make the choices relevant to the second step, usually devoted to marketing headquarters in conjunction with human factors specialists, hence for each interaction spotted, the list of all possible ways to solve the interaction problem is to be given.

References

[1] T. F. Bowen et al, The Feature Interaction Problem in Telecommunication Systems, Proceedings of the 7th SETS conference, July 1989.

[2] ETSI/NA 601-09, Service Life Cycle Reference Model

[3] ETSI/NA 611-01, Service and Service Feature Interaction: Service Creation, Service Management and Service Execution Aspects

[4] ETSI/NA1(93)03, Optional service features applicable to the new services

[5] EURESCOM P101, Pan European VPN, Volume 1, Service Specification

[6] I. Jacobson et al, Object-Oriented Software Engineering, A Use Case Driven Approach, Addison-Wesley, 1992.

[7] J. Rumbaugh et al, Object-Oriented Modelling and Design, Prentice Hall, 1991

[8] J. Muller et al, Perfection is not of this world: Debating a User-Driven Approach of Interaction in an Advanced Intelligent Network, TINA93, September 1993.

A Building Block Approach to Detecting and Resolving Feature Interactions

F. Joe LIN and Yow-Jian LIN
Bellcore, 445 South Street, NJ 07960, U.S.A.

Abstract. This paper presents a methodology we envision for detecting and resolving feature interactions. The methodology is based on a building block approach, in which features and their operating contexts are building blocks that can be composed in any combination to detect and resolve their interactions. This methodology is applicable to the phases in the software life cycle that address the creation of new features such as requirements, specification, and verification. By creating a well defined process for determining feature compatibility, with clearly defined steps and appropriate techniques/tools, it will then be possible to systematically model features, and detect and resolve interactions among features. The primary goal is to provide a support environment which feature designers can use to specify and verify the requirements of a feature, detect its possible interactions with other features, and finally verify the resolution of any detected interactions. As an ongoing effort at Bellcore, the paper also reports our current status of experiments with this methodology.

1. Introduction

An important step towards managing feature interactions is to ensure that the specification of a telecommunication feature is of high quality. That is, the specification should capture nothing but intended behaviors, and at the same time anticipate all possible influences from its operating environment, including the behaviors of users, network components, and other features. This paper presents a methodology for modeling feature logic and its operating environment as units of add-on functions, or building blocks, that can be combined for detecting incomplete or incorrect specifications. In particular, it addresses two fundamental issues: how to apply verification techniques to the feature-interaction problem; and how to structure the modeling of features to facilitate the verification of various combinations of features, or feature packages.

Detecting interactions can be viewed as a process of checking whether the actual behaviors of a feature, in the presence of other features, are different from the intended behaviors of the feature. In specifying a feature, one can describe it both as a sequence of computational steps (i.e., a low-level procedural specification) and as a set of functional properties (i.e., a high-level behavioral specification). The procedural specification represents the actual behaviors of a feature, and the behavioral specification captures the intended behaviors. For example, to specify an Originating Call Screening (OCS) feature, a procedural specification could specify that the call processing will first check the dialed number and then, if the number is not on the screening list, proceed to route the call (to its completion); whereas a behavioral specification could state that a caller with OCS cannot reach any number on the screening list. To detect interactions, one must determine if any scenario that is valid according to the procedural specification of a feature violates the behavioral specification of the same feature. In our OCS example, according to the procedural specification a user A with OCS can call B as long as B's number is not on A's screening list. However, if B has Call Forwarding, B may forward A's call to a number on A's screening list. Since the said scenario is not consistent with the behavioral specification of OCS, Call Forwarding and OCS interact.

The modeling of features requires a structured approach to address both the architectural issues and the dynamics of feature packages. Interactions can arise in a variety of ways[1] [2]. The modeling must allow an easy composition of feature specifications to facilitate the analysis of both single-user feature interactions as well as multiple-user feature interactions. Moreover, features operate in a range of contexts, i.e., the configurations of users and calls involved. Some features only affect one user, some affect both communicating parties of a single call, and some can change the behaviors of multiple related calls. The modeling must support the specification of various contexts and provide a simple mechanism for combining feature logics and contexts.

In our approach (see Section 2), a set of building blocks pave a structured approach to the modeling of features. For each feature, the procedural specification consists of two parts: the feature logic, representing the operations of the feature; and the corresponding context(s), representing the environment that the feature communicates with. A hierarchy of blocks constitutes the modeling of feature contexts. At the lowest level, the originating and terminating basic call models are the basis. Together with the blocks that represent all possible originating user behaviors, terminating user behaviors, and system/network behaviors, this approach provides three kinds of *basic feature contexts*: originating, terminating, and two-party. One can then create a specific context for the operation of a feature by combining instances of these basic feature contexts together. The feature contexts also facilitate the composition of feature specifications; it becomes a process of merging corresponding building blocks in their contexts, and then linking their logics to the resulting combined contexts.

Detecting interactions in this approach is a matter of checking whether the combined behavioral specifications and procedural specifications are consistent (see Section 3). In addition to a procedural specification, the modeling of each feature also provides a behavioral specification, expressed as a set of temporal logic formulas. These formulas represent, in the view of each feature, the properties that must be valid in all circumstances. Given a set of features, the conjunction of their temporal logic formulas forms their combined behavioral specifications. With a powerful verification tool, our work has created an environment that allows feature designers to specify and verify the behavioral specifications of each feature systematically.

This paper is organized as follows. Section 2 presents the concepts of our building block approach, outlining the steps taken to model features, compose feature packages, and incorporate resolutions. Section 3 briefly describes the verification process for detecting interactions, and discusses the tools required to support the modeling and verification steps based on our building block approach. Section 4 gives a status report of our ongoing research project, including the description of our specification and verification environment, building blocks, and tools for creating feature contexts; it also briefly explains the modeling of a sample feature. Section 5 follows with some concluding remarks.

2. A Building Block Approach

To detect feature interactions and verify their resolutions, we have developed a process for modeling features and feature packages, i.e., combinations of features. The process consists of five major steps:

1. Define Basic Call Models (BCMs)
2. Derive Basic Feature Contexts (BFCs) from BCMs

3. Specify features in terms of BFCs
4. Compose feature specifications for detecting interactions
5. Model interaction resolutions for verifying their correctness

In this approach, the specifications of individual features are building blocks for creating the specifications of feature packages subject to interaction detection. In addition, the specification of each feature requires a model of its own *feature context*. Steps 1 and 2 create the basic building blocks that allow feature designers to build various feature contexts, facilitating the modeling of individual features in Step 3. Step 4 provides the basis for detecting interactions, whereas Step 5 is for verifying the correctness of resolutions. This section describes the basic concepts of each step. Some of the steps, including the creation of BFCs and the modeling of features will be discussed in detail in Section 4.

2.1 Define Basic Call Models (BCMs)

Telecommunications features are add-on functions over a basic phone service. In order to model the operations of a feature, an essential step is to model the basic phone service, i.e., to define BCMs. BCMs represent the application-layer protocols at the user-network interface of telephone switching systems; they may vary depending on the kinds of services offered and the structure of switching software. For narrowband voice services, we follow the BCMs defined in Bellcore Release 0.2 switch generic requirements[3]. As shown in Appendices A and B, there are two BCMs: one for call origination, called *Originating BCM*; and the other for call termination, called *Terminating BCM*. The states in these BCMs are points in call (PICs), indicating the various stages that a call progresses through until its completion. In addition to PICs, the BCMs also define Detection Points (DPs) at which call processing control along with call-associated data can be transferred from the BCMs to a feature, if so desired. This allows a feature to add new logic to the original protocols in order to provide new services.

2.2 Derive Basic Feature Contexts (BFCs) from BCMs

Although features communicate directly with BCMs, their behaviors also depend upon the environment they are operating on. Therefore, a model of both the BCMs and the environment, called a *feature context*, is needed to fully describe the operations and properties of a feature. Some features affect only one side of the call processing, either the originating side or the terminating side; hence their feature contexts involve the user behavior and the BCM in the corresponding side, as well as the system behaviors perceived from the same side. Others concern end-to-end call processing, and require a feature context that involves users of both ends. While there are many different feature contexts, we find that it is sufficient to construct these contexts from three *Basic Feature Contexts* shown in Table 1:

1. Originating BFC - This BFC models the feature context that involves only the user in the originating side.
2. Terminating BFC - This BFC models the feature context that involves only the user in the terminating side.
3. Two-Party BFC - This BFC models the feature context that involves both the originating side and the terminating side users.

Table 1. Building Blocks for Feature Contexts

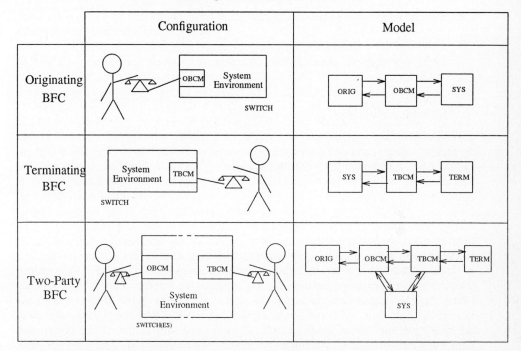

The left column of Table 1 shows the physical configuration of a particular BFC; the right column shows its corresponding model as a network of communicating processes, depicted by blocks and arrows. Blocks ORIG and TERM model the originating and the terminating users' behavior, respectively. Block OBCM is for the originating BCM, and block TBCM is for the terminating BCM. Block SYS represents an abstraction of the system environment that BCMs reside in: in originating BFC and terminating BFC, SYS models how the environment may interact with OBCM and TBCM, respectively; and in two-party BFC, with both. The arrows between the blocks indicate the communication channels between the processes.

Note that a two-party BFC is effectively the composition of an originating BFC and a terminating BFC. It is possible to construct feature contexts entirely based on two-party BFCs, or on combinations of originating BFCs and terminating BFCs. Nevertheless, we choose to have all three of them: it is less complex to use an originating BFC or a terminating BFC to construct a feature context that involves only one user; and it is convenient to have a two-party BFC ready for constructing feature contexts that involve more than one user. The two-party BFC is also useful for expressing the composition of multiple features specified in terms of only originating and terminating BFCs.

2.3 Model Features in terms of BFCs

Once BFCs are available, the next step is to piece together proper BFCs with a feature's logic to model the operations of the feature. Figure 1 illustrates how four typical features are modeled. First, a brief description of each feature from its users' perspective:

1. Originating Call Screening[4] - The user defines a screening list of telephone numbers; all outgoing calls to those number will be blocked.

2. Denied Termination[5] - The user denies the termination of all incoming calls.

3. Call Waiting[6] - The user will be notified of an incoming call by a tone burst when he/she is already on another call. The user can then switch back and forth between the two calls by flashing the hook so that one call is on hold while the other is active. If the user hangs up one call while the other call is still on hold, he/she will be rung back.

4. Call Forwarding on Busy[7] - The user can have other incoming calls forwarded to a different number when he/she is already on a call.

To model a feature in terms of its context, one has to identify a proper set of BFCs that the feature requires, and at the same time to determine the BCMs in these BFCs that the feature has to communicate with through their PICs. The required BFCs and the interacting BCMs for the above features are shown in Figure 1, with the dashed lines connecting each feature's logic with its interacting BCMs. As the figure indicated, Originating Call Screening requires only an originating BFC as its feature context. Likewise, the feature context for Denied Termination only needs a terminating BFC. Call Waiting has two possible feature contexts depending on whether the first call is an incoming or an outgoing call. In the former, the feature context requires two terminating BFCs; in the latter, it requires one originating BFC and one terminating BFC. The feature context for Call Forwarding on Busy is a bit complicated; it requires a two-party BFC to model the arrival of a new incoming call and a second two-party BFC to model the forwarding of the call. After the forwarding has taken place, the OBCM of the first two-party BFC and the TBCM of the second two-party BFC will be directly connected and form a new two-party BFC. Note that Call Forwarding on Busy only communicates with the TBCM of the first BFC and the OBCM of the second BFC.

It is possible that, while composing BFCs together to form a feature's context, one may need to merge some building blocks. In the first feature context of our Call Waiting example, the block encapsulating two TERMs indicates that the two TERMs, originally in two independent BFCs, are modeling the same user and should now be merged. Similarly, we merge ORIG and TERM in the second feature context of Call Waiting.

2.4 Compose Feature Specifications for Detecting Interactions

To detect interactions among features, the specifications of the features need to be brought into a single context. In our building block approach, composing feature specifications is not much different from modeling individual features in terms of BFCs. Using the feature contexts of individual features as the building blocks, our approach first forms a composite context for all the features involved. It then consolidates and attaches all the feature logics to this composite context. During the composition, we identify (manually at this time) and merge different feature contexts to yield a concise composed context.

With feature contexts and feature logics, the composition in our approach deals uniformly with features of a single user as well as features of multiple users. Based on the feature specifications in Figure 1, Figures 2 and 3 show two compositions of features belonging to a single user: Originating Call Screening, Denied Termination, and Call Waiting in Figure 2; and Call Waiting and Call Forwarding on Busy in Figure 3. On the other hand, Figure 4 is a composition of features belonging to two users: one has Denied Termination and the other has both Call Waiting and Originating Call Screening.

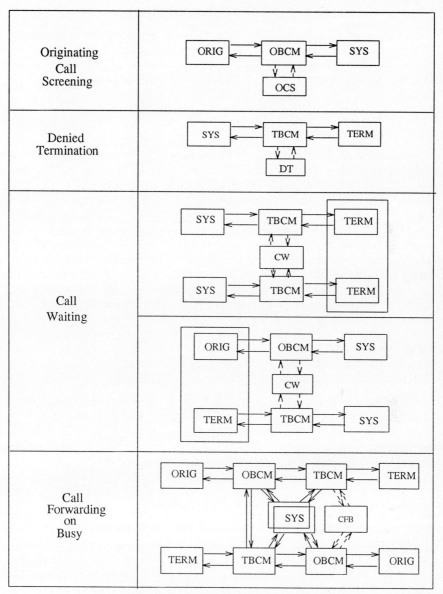

Figure 1. Call Contexts for Four Features

Details of how to detect feature interactions in those composite models are discussed in Section 3.

2.5 Model Interaction Resolutions for Verifying Their Correctness

It is necessary to ensure that any proposed resolution for a detected interaction indeed resolves the interactions. To model resolutions for verifying their correctness, our approach follows a distributed feature management concept[8]. Instead of modifying individual feature logics to resolve interactions, a management agent, called *feature*

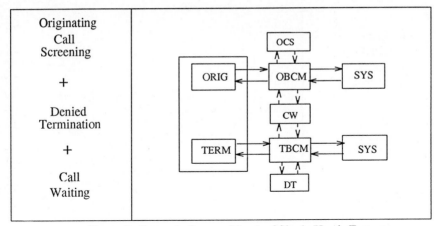

Figure 2. Example Composition 1 of Single User's Features

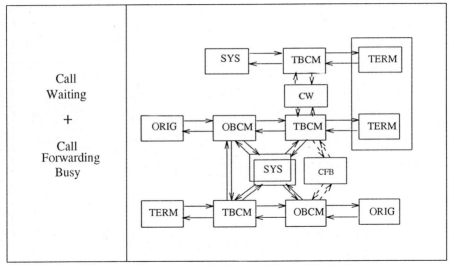

Figure 3. Example Composition 2 of Single User's Features

manager, is responsible for resolving interactions among features of a single user and for coordinating with other feature managers to resolve interactions among features of multiple users. Figure 5 illustrates how the building block approach can easily incorporate a new feature manager entity to model the resolutions needed in Figure 2. All communications between BCMs and features must now go through the feature manager, but the feature logics and the BCMs remain the same. The feature manager acts as an arbitrator between BCMs and features, and enforces the (manually derived) resolution scheme over all the features. One can then re-analyze the new model to confirm that the resolution is correct and that no additional interactions exist.

3. Tools Required to Support the Process

This section discusses tools required to support the process described in the last section. These tools can be divided into four major categories:

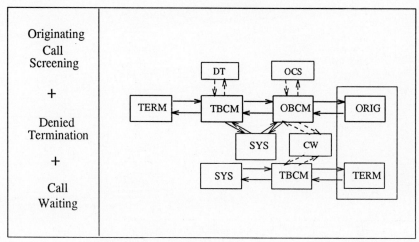

Figure 4. Example Composition of Multiple Users' Features

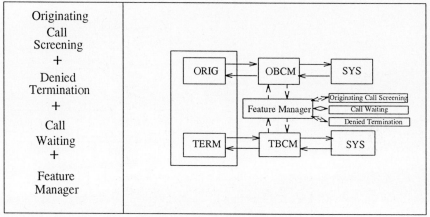

Figure 5. Feature Manager of Features

1. Specification tools
2. Verification tools
3. Composition tools
4. Animation tools

Each of them is discussed in detail in the following.

3.1 Specification Tools

This process uses two specification tools: one is for the low-level procedural specification, describing in procedural steps *how* a feature and its context should operate; the other is for the high-level behavioral specification, describing in logic assertions *what* properties a feature and its context must exhibit. Such a two-level specification approach is applied throughout the entire process; it is the key to being able to detect and resolve a variety of feature interactions.

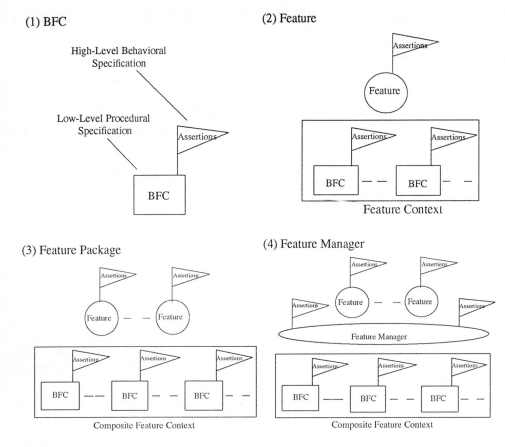

Figure 6. Two-Level Specifications Applied in the Entire Process

Starting from the BFCs, each building block is specified in two-level specifications. The BFCs are composed to form a feature context. With the feature context, the feature is then specified in two-level specifications as well. Features are further composed to form a feature package. In this process (see Fig. 6), as smaller building blocks are composed into a larger one, their low level procedural specifications are merged into one model. If there are no interactions, the properties of the composition of the smaller blocks should be the conjunction of their individual properties. Otherwise, individual properties of certain building blocks will be violated in the composite model. The improper interactions among features in a feature package can thus be detected. If so, a feature manager, also in two-level specifications, can be specified and inserted into the model for resolution verification. Using such an approach detecting and resolving feature interactions is converted into a verification problem. In each step of Fig. 6, the merged low-level procedural specifications (indicated by circles, ovals, and rectangles) must be verified against all the high-level behavioral specifications (indicated by flags).

For example, there is an interaction between Call Waiting (CW) and Denied Termination (DT) in Fig. 2. This is detected by verification as the conjunction of the following two individual assertions becomes the property of the composed model.

— Denied Termination's Assertion - *No incoming call is allowed to terminate at a line with DT activated.*

— Call Waiting's Assertion - *Call termination at an idle line with CW service should be unaffected.*

Next, more details at each level of specification are discussed.

3.1.1 Low-Level Procedural Specification ESTELLE[9], LOTOS[9], SDL[9], and Promela[10] are examples of the languages at this level, which must allow us to easily

— specify the protocols such as POTS and ISDN, as well as the add-on features built upon these protocols.

— construct many varieties of complex models (such as the ones in Figs. 1-5) from a few simple basic models (building blocks in Table 1).

As a result, the important characteristics of a language at this level are:

- It must support both typed data and parameterized messages. All the user-network interface protocols such as POTS and ISDN are rich in both typed data and parameterized messages.
- It must support communications via both message passing and shared data. The former is convenient for specifying the communications between protocol entities; the latter, for specifying the communications between protocols and their add-on features.
- It must support nondeterminism. Though OBCM, TBCM, and features can be specified as deterministic processes, ORIG, TERM, and SYS which model the subscriber's behavior and the switching system environment rely heavily on nondeterminism to form a closed model.
- It must support process instantiation. This allows us to create complex models from a few process modules such as TERM, ORIG, SYS, OBCM, and TBCM.
- It must support binding of channels to processes at the time of process instantiation. This allows us to connect multiple process instances into a sophisticated model.
- It must support assignment of shared data to processes at the time of process instantiation. This allows us to make provision for accommodating features on top of each BCM.

Almost all the characteristics listed above can be found in many of formal specification languages today such as ESTELLE, LOTOS, SDL, and Promela.

3.1.2 High-Level Behavioral Specification Our high-level behavioral specification language is based on temporal logic. Temporal logic has been proved well-suited to describing the properties of concurrent models such as the one we are dealing with here[11]. Moreover, among variations of temporal logic, we found the three basic linear time temporal properties as defined in [12] sufficiently powerful in declaring all the properties of telecommunication features. These properties are invariance, response, and precedence, represented by the temporal formulas below:

invariance	always t
response	t1 implies eventually t2
precedence	t1 implies (t2 until t3)

where 't's can be temporal or propositional formulas.

Furthermore, our experiments show that all the properties we need to express so far fall into the following subset of formulas:

$$t ::= $$
$$\quad \text{t1 implies eventually p } |$$
$$\quad \text{t1 implies (p until false) } |$$
$$\quad \text{always !t1 } |$$
$$\quad \text{always p.}$$

$$t1 ::=$$
$$\quad \text{t1 implies (!p2 until p1) } |$$
$$\quad \text{t1 implies eventually p } |$$
$$\quad \text{p.}$$

where 'p's are propositional formulas and '!' is logic operator *not*.

To make the temporal ordering expressed by these formulas even clearer, we create the following shorthand notations for writing the assertions for BFCs, features, and feature managers.

Temporal Formula	Shorthand Notations
t1 implies (!p2 until p1)	t1 FOLLOWED BY p1 WITHOUT PRECEDING p2
t1 implies eventually p	t1 FOLLOWED BY p (if derived from t1)
t1 implies eventually p	t1 IMPLIES EVENTUALLY p (if derived from t)
t1 implies (p until false)	t1 IMPLIES FOREVER p

In other words, all properties belong to either of the following:

- the fact that a sequence of events must lead to a certain consequence
- the invariant

The sequence of events is expressed by temporal formula t1, which can be as simple as an event denoted by a propositional formula, or as complex as one event after another with or without preemption defined recursively. An example for the latter is p1 FOLLOWED BY p2 FOLLOWED BY p3 WITHOUT PRECEDING p4 (interpreted as [p1 FOLLOWED BY p2] FOLLOWED BY p3 WITHOUT PRECEDING p4 according to the above definition.) This defines a sequence of events depicted as follows.

$$P1 \quad\quad P2 \quad\quad P3$$
$$\bigcirc\!\!\rightarrow\!\!\bigcirc\cdots\rightarrow\!\!\bigcirc\!\!\rightarrow\!\!\bigcirc\cdots\rightarrow\!\!\bigcirc$$
$$\text{no P4}$$

This is a sequence of events in which p1, p2, and p3 occur in order and p4 does not occur between p2 and p3 (p4 is a preemption.) There may be other events between p1 and p2 or between p2 and p3. But we are not interested in them.

The consequence is expressed by either operator IMPLIES EVENTUALLY or IMPLIES FOREVER: the former asserts that something is going to happen eventually; the latter something always holds henceforth. The invariant is claimed by the ALWAYS (always) operator. Section 4 will show some specification examples using these temporal notations.

3.2 Verification Tools

Verification is the core mechanism in our approach. To apply verification effectively in our process, the following characteristics of the tools are needed:

- A computer-aided system rather than a manual proof system. Considering the complexity of the problems it is infeasible to use any manual proof system as the verification engine for all levels of verification. The engine needs to be a computer-aided system that requires as little human effort as possible in carrying out the verification.

- More geared to protocol verification. Detecting feature interactions is similar to protocol verification problems. The BFCs can be considered as the application layer protocols that offer basic services. The features built upon BFCs can be viewed as the additions or the enhancements to these protocols in order to offer advanced services. Consequently, we need tools that are designed with characteristics and nature of communication protocols in mind.

- Efficient verification of temporal logic assertions. As explained before, temporal logic assertions capture the properties of building blocks that can be carried forward as they are composed to form larger models. Verification of those assertions is the key to detecting interactions among features. Thus, such a capability is needed in the verification tools and must be implemented with an efficient verification algorithm.

- Scale up to handle a state space of 100 million states or more. Our composed model can grow to tremendous complexity from a few basic building blocks. As a result, the desired verification tools must be capable of handling a model of 100 million states or more.

- Scale up to handle hundreds of temporal logic assertions automatically. Temporal logic assertions are the center of our approach. As complex models are composed from simple ones, the number of assertions can easily reach several hundreds. Tools must allow us to verify hundreds of assertions without any human intervention.

- Strong support in debugging violations. Because interaction detection is equivalent to assertion violation of composed features, the tools must provide adequate and detailed information on how the violation occurred so that we can track down the source of interaction. The information could also provide us clues on how to resolve the interaction by inserting an appropriate feature manager in the model.

In summary, we are mainly interested in efficient model-checking tools that can sustain the complexity of the problems we are dealing with here. Many such tools based on theoretical foundation of temporal logic have been created in the past few years. Examples are COSPAN[13], SMV[14] [15], and SPIN[10].

3.3 Composition Tools

In this process, two levels of composition are considered:

1. Composition of a feature context for feature specification. A feature context can be composed from BFCs. During such a composition, the following information is crucial:

    — What kinds of BFCs, as well as how many for each kind, are required

    — The BCMs in the feature context that the feature needs to interact with

— The DPs (Detection Points) in the BCMs where the feature needs to add new logic to call processing

The composition tool at this level would take the above information and automatically compose the feature context required for a particular feature.

2. Composition of multiple features for detecting and resolving their interactions. After a feature is formally specified and verified, all its information including its two-level specifications, its context, and its interacting BCMs and DPs will be available for access. The composition of multiple features into a single model involves a series of questions:

— How many subscribers are involved in these features?

— Which subscribes own which features?

— What features are specified with multiple BFCs and belong to the *same* subscriber? Incrementally compose each group of them by asking the following questions:

— What BFCs can be shared by those features?

— Which BCMs in those BFCs are the joint points?

— Finally, what features are specified with single BFC? Compose them to the model based on which subscribers they belong to.

Through this series of questions the composition tool at this level should be able to derive the composite model from individual feature's information and under the user's guidance. We hope that the composition at this level can be automated to its greatest extent.

3.4 Animation Tools

Due to the complexity of the problem domain, two animation tools are well-justified:

1. Animation of the models composed during the process. This would create a much more user-friendly environment by depicting the process described in this paper.

2. Animation of the interaction trace. This would allow much quicker user response in debugging the improper interactions as well as allow lively animation of how the processes in a complex model interact with one another to fulfill a service requirement.

4. Status of Our Experiments at Bellcore

We are currently experimenting with this feature interaction detection and resolution process at Bellcore. Two major tasks are involved in these experiments: the creation of building blocks, and the development of supporting tools. Both tasks are closely interrelated; each depends on the other for making real progress. On the one hand, we need the building blocks to experiment with the appropriateness and effectiveness of the tools. On the other hand, the tools speed up the creation of the building blocks for the next phase of experiment. Therefore, we mix the discussions of both tasks in the subsequent sections roughly following their chronological order in our ongoing effort.

4.1 Specification and Verification Tools

Among many possible choices of specification and verification tools, we selected the verification tool SPIN and its specification language Promela[10] because of two reasons: first, Promela and SPIN meet all our requirements of the specification and verification tools (Secs. 3.1 and 3.2) except in two areas; second, SPIN is stable, easily available and well-documented.

The two exceptions we identified are

1. No temporal logic assertions are accepted by SPIN directly. They need to be translated into Promela "never-claims" for verification.

2. SPIN is not able to handle hundreds of temporal logic assertions automatically. The best way of running SPIN is to verify one assertion at a time.

We are able to enhance Promela and SPIN by building our own tools on top of them. These enhancements are discussed here. They are all in the area of temporal logic:

1. Specification. We have done two things. First, although SPIN has the power to support the verification of any linear time temporal logic assertions[16], it doesn't accept temporal logic assertions directly. A temporal logic assertion has to be translated into a Promela "never-claim", before any verification can be executed. We have built a tool that hides such a translation from users so that all reasoning can be done at the level of temporal logic formulas (those defined in Section 3.1.2) rather than at that of Promela "never-claims".

   Second, in Promela there is no primitive to state the fact that a specific message has just been received from a specific channel. This kind of proposition occurs frequently in dealing with features. We have built a tool to add one such primitive to the Promela propositional formulas, which allows us to assert the receiving of a message from a channel without any manual effort on the specification.

2. Verification. Temporal logic assertions are the center of our approach. As we compose complex models from simple ones, the quantity of them can easily reach several hundreds. Dealing with so many temporal logic assertions is a very tedious, if not infeasible, task due to the following reasons:

   — The best way of verifying assertions using SPIN is one at a time. By focusing on one assertion, SPIN gives the best state space coverage when dealing with complex models. Also, it is much easier to identify which assertions didn't pass the verification.

   — For *P IMPLIES EVENTUALLY Q* and *P IMPLIES FOREVER Q* types of assertions, we have to prove that *ALWAYS !P* is violated in the model before launching the verification. Without doing so, both types of assertions may be vacuously true because 'P's are always false in the model.

   — If an assertion passes the verification, we would like to generate a call scenario that manifests the property specified by the assertion. This information can help us better understand the interaction among features.

   What SPIN provides is an efficient and robust verification engine. To effectively use this engine, we have built an automated verification environment called WHEEL†. WHEEL incorporates into its environment the enhancement tools we built for the temporal logic specification. It automates the whole verification procedure as depicted in Fig. 7.

In this environment, the specifier doesn't have to deal with Promela "never-claims". The specifier would write temporal logic assertions and store them in a database. Then with one command, WHEEL would retrieve from the database the assertions yet to be verified, convert them into Promela "never-claims", and dispatch SPIN to verify each assertion from one of the following three levels.

Level 1 Existence verification. Done for *P IMPLIES EVENTUALLY Q* and *P IMPLIES FOREVER Q* types of assertions. We prove that *ALWAYS !P* is false (i.e. P exists) at this level. If this passes, both types of assertion will move to Level 2.

Level 2 Consistency verification. Done for all assertions. We prove that no assertions are violated in the model. All assertions except the *ALWAYS* ones will move to Level 3 if they pass this level of verification. The *ALWAYS* assertions require no scenario generation since they hold in all scenarios.

Level 3 Scenario generation. Done only for *P IMPLIES EVENTUALLY Q* and *P IMPLIES FOREVER Q* types of assertions. The purpose is to produce a call scenario that would demonstrate the property specified by an assertion. This can be accomplished by verifying *P IMPLIES FOREVER !Q* for *P IMPLIES EVENTUALLY Q*, and *P IMPLIES EVENTUALLY !Q* for *P IMPLIES FOREVER Q*.

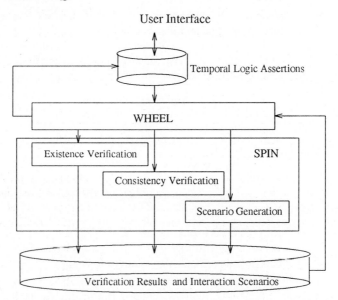

Figure 7. WHEEL - An Automated Verification Environment Based on SPIN

---

† WHEEL is named after the fact that it runs SPIN one round after another until all required verifications are executed.

The status of each assertion with respect to the level of verification will be tracked in the assertion database. WHEEL decides which level of verification to run next based on this status.

Each SPIN execution stores the verification result in a result database. WHEEL checks the result of each verification level and updates the status of the corresponding assertion in the assertion database accordingly. A temporal logic assertion, under the management of WHEEL, can thus move down the pipelined procedure of existence verification, consistency verification, and/or scenario generation without any human intervention. If any unexpected result occurs, WHEEL will instruct SPIN to generate an error trace and store it in the result database for later debugging. WHEEL gives us a powerful environment to deal with any large quantity of temporal logic assertions. It also hides the low-level details of how to run SPIN from the users of the process.

4.2 Creation of Building Block BFCs

We have created the two-level specifications for the three BFCs defined in Sec. 2, and verified their correctness using the SPIN/WHEEL. We wrote the two-level specifications of terminating BFC and originating BFC manually but were able to semi-automatically compose the two-level specifications of two-party BFC from those of both originating and terminating BFCs.

Table 3 below gives the complexity of each model in terms of the sizes of their state space and state vector, and the sizes of their Promela specifications and temporal logic assertions. All these models were verified using SPIN's exhaustive search[10].

Table 3. Complexity of BFCs (All Data Rounded)

Type of BFC	Size of State Space in Reachable States	Size of State Vector in Bytes	Number of Lines in Promela Specification	Number of Assertions in Temporal Logic
Originating BFC	15k	60	600	20
Terminating BFC	6k	90	450	20
Two-Party BFC	340k	160	900	40

To give readers a better picture, the complete two-level specifications of one of the BFCs, the terminating BFC, were given in Appendix C.

Since the BFCs are the building blocks at the lowest level, their complexities determine the complexities of all the models that are to be built upon them. As exemplified by the specifications in Appendix C, we have structured the specifications of all the BFCs based on the idea of *options*. Through options, the state space of a BFC is decomposed into a number of disjoint subspaces. If all options are turned on, verification will exercise the complete state space, i.e., every branch in BCMs will be traversed. If all options are turned off, verification will exercise only the most basic call scenario, which is an execution flow of PICs 1-2-3-4-5-6-7-8-9-1 for the Originating BCM in Appendix A or an execution flow of PICs 11-12-13-14-15-16-11 for the Terminating BCM in Appendix B. Between these two extremes, we can turn any options on or off to control which portions of the state space should be exercised by the verification. These options give us a finer control over the complexity of the model. This can be used in two ways:

1. It makes incremental specification of a feature possible. A feature is specified with respect to its feature context, composed from the BFCs. With BFCs' options the complexity of the feature context against which a feature will be specified can be determined. We thus can specify a feature incrementally by first considering the

operations of the feature in the simple contexts, and then gradually move to the ones in the complex contexts.

2. It helps cope with the state explosion problem. Our feature model may grow into such a complexity that eventually no exhaustive but supertrace search is possible. Furthermore, even verification using supertrace may reach its limits because of both resource and schedule restraints. BFCs' options allow us an alternative way of coping with this state explosion problem by focusing on certain important portions of the state space in one verification[17].

We have defined one set of options for each of the originating and the terminating BFCs. The two-party BFC then uses both sets as its options. As an example, Table 4 below lists all the options of the terminating BFC and also shows how different settings of these options affect the verification complexity of the terminating BFC.

Table 4. Complexity of Terminating BFC by Options

OPTIONS	Complexity based on Exhaustive Search							
	CASE						
	1	2	3	4	5	6	7
UNSUC_SETUP	off	off	off	off	off	off	on
T_EARLY_RELEASE	off	off	off	on	on	on	on
SS7_FAILURE	off	off	off	off	off	off	on
T_CALL_REJECTED	off	off	off	off	off	off	on
NO_ANSWER	off	off	on	on	on	on	on
T_IMMD_ANSWER	off	off	off	off	on	on	on
T_BKWD_RELEASE	on	on	on	on	on	on	on
T_REANSWER	off	off	off	off	off	on	on
LINE_BUSY	off	on	on	on	on	on	on
Size of State Space in Reachable States	380	468	666	1485	1640	5245	6242

Readers should notice that from one extreme (Case 7) to another extreme (Case 1) is a reduction of almost 20:1 in the size of state space.

4.3 Composition Tools for Producing Feature Contexts

We have built an automated environment that can produce the feature context required for a feature logic without any manual effort. This environment is depicted in Fig. 8. Internal to the environment are a BFC repository and two automated tools.

The repository keeps the specifications of three types of BFCs. The first tool is used to compose together all the BFCs required for a feature logic. The tool requires the user to supply a specification that states what kinds of BFCs, as well as how many for each kind, are required for the composition. Example specifications of this kind for the features in Fig.1 are illustrated in Table 5.

Note that an originating BFC is represented by an "O"; a terminating BFC is represented by a "T"; a two-party BFC by an "<OT>". The repetition of a BFC can be represented by a number in front of it. For example, TT can be represented by 2T.

The second tool is used to generate a feature specification template, which can later be used to fill in details of the feature logic, as well as the hooks between this feature template

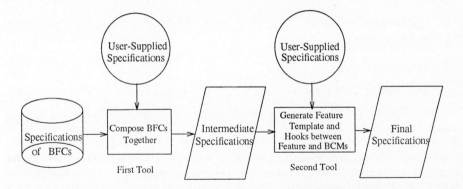

Figure 8. An Automated Environment for Producing Feature Contexts

Table 5. User-Supplied Specifications for the First Tool

Features	Feature Contexts
CALL FORWARDING BUSY LINE	2<OT>
CALL WAITING	2T
CALL WAITING	OT
ORIGINATING CALL SCREENING	O
DENIED TERMINATION	T

and the BCMs. This tool requires the user to supply two pieces of information:

1. The BCMs in the feature context that the feature needs to interact with
2. The DPs in the BCMs where the feature needs to add new logic to the call processing

Example specifications of this kind are illustrated in Table 6 for the Call Waiting feature.

Table 6. User-Supplied Specifications for the Second Tool

Features	Feature Contexts	BCMs and DPs
CALL WAITING	2T	(1,<t_mid_call,t_cleared,t_release_timeout>), (2,<t_busy,call_accepted,t_mid_call,t_cleared,t_release_timeout>)
CALL WAITING	OT	(1,<o_mid_call,o_disconnected,o_release_timeout>), (2,<t_busy,call_accepted,t_mid_call,t_cleared,t_release_timeout>)

Note that each specification consists of a list of binary tuples. The first parameter of the tuple gives the identity of a BCM, whereas the second parameter gives a list of DPs in the BCM.

Any feature context generated from this automated environment is error-free and preserves all the properties of the composed BFCs. The feature specifier can thus always begin with a clean feature context, and then work on top of it to create the feature specification.

4.4 Creation of Building Block Features

At the time of writing this paper, our major effort lies in this area. We are creating the specifications of the following features in preparation for our next phase of experiment.

1. Call Waiting[6]
2. Call Forwarding on Busy[7]
3. Three Way Calling[18]
4. Call Waiting Deluxe[19]

We will use the Call Waiting in the 2T context to briefly explain how to create a feature on top of its context. The detailed model of Call Waiting in the 2T context is depicted in Fig. 9. Note that the feature context automatically created for Call Waiting has a unique identity assigned to each of its BCMs and BFCs. Two terminating BFCs were identified as ONE and TWO, respectively. Two terminating BCMs were identified as FIRST and SECOND, respectively. Through arrays of data and process instantiation, each BFC has its own shared variables, channels, and processes. A feature template was automatically generated as well. This template contains all the pre-feature call processing logic of the Call Waiting's DPs.

Now to illustrate how to specify Call Waiting on top of its feature context, we give the specification of the following property of Call Waiting as an example. A complete list of Call Waiting properties can be found in [6].

If a call terminates at a line with the CW service that is at a stable call state, then the calling party should hear audible ringing and the called party should hear the CW tone.

First, we specify the feature's property in temporal logic. All propositions in the property are identified and defined in the following.

- call terminates ≡ tsys_forward[TWO]?[setup]
- line with the CW service ≡ feature[CALL_WAITING]
- a stable state ≡ (_state[FIRST] == T_ALERTING || _state[FIRST] == ACTIVE)

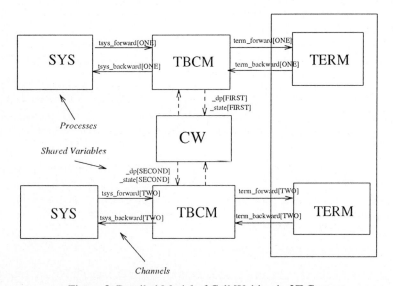

Figure 9. Detailed Model of Call Waiting in 2T Context

- caller hears audible ringing ≡ tsys_backward[TWO]?[setup_cmp]
- called party hears the CW tone ≡ term_forward[ONE]?[tone]

The temporal logic operator *IMPLIES EVENTUALLY* is then used to express the whole property in two assertions.

(tsys_forward[TWO]?[setup] && feature[CALL_WAITING] &&
(_state[FIRST] == T_ALERTING || _state[FIRST] == ACTIVE))
IMPLIES EVENTUALLY
(tsys_backward[TWO]?setup_cmp)

(tsys_forward[TWO]?[setup] && feature[CALL_WAITING] &&
(_state[FIRST] == T_ALERTING || _state[FIRST] == ACTIVE))
IMPLIES EVENTUALLY
(term_forward[ONE]?[tone])

Second, we procedurally specify how the feature operates to satisfy the property. We identify that the feature has to take actions at the DPs T_BUSY and CALL_ACCEPTED of the second BCM. Using the feature template, which is automatically generated using the environment described in Sec.4.3, we modify the call processing of detecting a DP T_BUSY in the second BCM from the left column to the right column of the following table.

Pre-Feature Logic in Feature Template	Call Waiting's Feature Logic

atomic
{
  _dp[second_leg] == T_BUSY ->
    _dp[second_leg] = NULL;
  if
  :: features[CALL_WAITING] ->
    tsys_backward[context_two]!sub_busy;
    _state[second_leg] = END_DONE
  :: !features[CALL_WAITING] ->
    tsys_backward[context_two]!sub_busy;
    _state[second_leg] = END_DONE
  fi
}
	atomic
	{
	  _dp[second_leg] == T_BUSY ->
	    _dp[second_leg] = NULL;
	  if
	  :: features[CALL_WAITING] ->
	    term_forward[context_two]!setup;
	    _state[second_leg] = PRESENT_CALL
	  :: !features[CALL_WAITING] ->
	    tsys_backward[context_two]!sub_busy;
	    _state[second_leg] = END_DONE
	  fi
	}

Likewise, we modify the call processing of detecting a DP CALL_ACCEPTED in the second BCM as illustrated in the table below.

Pre-Feature Logic in Feature Template	Call Waiting's Feature Logic

atomic
{
  _dp[second_leg] == CALL_ACCEPTED ->
    _dp[second_leg] = NULL;
  if
  :: features[CALL_WAITING] ->
    term_forward[context_two]!ring;
    tsys_backward[context_two]!setup_cmp;
    _state[second_leg] = T_ALERTING
  :: !features[CALL_WAITING] ->
    term_forward[context_two]!ring;
    tsys_backward[context_two]!setup_cmp;
    _state[second_leg] = T_ALERTING
  fi
}
	atomic
	{
	  _dp[second_leg] == CALL_ACCEPTED ->
	    _dp[second_leg] = NULL;
	  if
	  :: features[CALL_WAITING] ->
	    term_forward[context_two]!tone;
	    tsys_backward[context_two]!setup_cmp;
	    _state[second_leg] = T_ALERTING
	  :: !features[CALL_WAITING] ->
	    term_forward[context_two]!ring;
	    tsys_backward[context_two]!setup_cmp;
	    _state[second_leg] = T_ALERTING
	  fi
	}

The final specifications of the Call Waiting in the 2T context consist of 40 temporal assertions and 170 lines of Promela specification. Remarkably, these specifications were able to cover nearly all the informal feature requirements as stated in the original feature specification document[6].

In addition to specifying the feature itself, two extra tasks are needed to create a complete model. One is to specify the Call Waiting subscriber's behavior by merging two TERM processes; the other is to modify the "init" process to accommodate this merging. We skip the details of both due to the space limit.

4.5 Composition Tools for Detecting Feature Interaction

We have been developing the conceptual ideas and the algorithms for this set of tools. We envision that these tools (See Fig. 10) would take user-supplied specifications, retrieve feature specifications and all their associated information such as feature contexts and interacting BCMs and DPs, and then compose the specifications that accommodate all features in a single model. Such specifications can be input to SPIN/WHEEL for detecting feature interactions. If any interactions are detected, the animation tools will give users a picture of what has gone wrong. Much work remains to be done in this area.

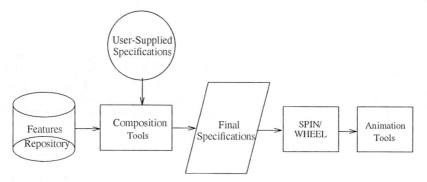

Figure 10. Composition Tools for Detecting Feature Interactions

5. Concluding Remarks

One of the major problems in managing feature interactions has been lack of a systematic approach to identify the interactions automatically. This paper presents a new approach, with clearly defined steps and appropriate analysis techniques, to facilitate the creation and use of computer-aided tools to model, configure, analyze, and store the information necessary to deal with the feature interaction problem. The building blocks, the modeling of feature contexts, the two-level specifications, and the verification of logic assertions are key concepts in our approach to detecting feature interactions early in the feature deployment process. Several other research organizations are taking similar approaches[20] [21] [22]. Some of the differences have been the choices of specification languages and the ways features are specified and composed. Note that the proposed modeling approach is not dependent on the languages used; however, the tools that support the languages could determine the effectiveness of the approach.

An important step to fully explore all potential feature behaviors is to identify a necessary and sufficient feature context for the features under analysis. While there are heuristics available, a general mechanism of generating contexts for any feature or feature packages is unknown at this time. Our choices of BFCs and the building block approach provide a nice environment for helping human experts experiment and create appropriate feature contexts.

Though the communications between BCMs and features are now modeled by data sharing, they may also take the form of message passing. It appears that the automatic conversion of the data sharing mechanism to the message passing mechanism is feasible. With such a conversion in place, we expect to deal with both switch-based (non-AIN) and AIN features uniformly using the proposed approach.

The amount of effort to create a working environment for modeling features and detecting interactions is substantial. It requires a deep understanding of the existing switching software and features, a fair amount of knowledge about formal techniques, a lot of experience in using and building modeling and verification tools, and, not to be taken lightly, a tedious process of formally specifying every feature. This paper describes some of the complexities in developing tools for modeling features and verifying correctness of specifications. It also illustrates how to use these tools to specify features and conduct verification. Nevertheless, there is still a long way to go to have most features specified. With the composition tools that we have developed and some more that we intend to build, we welcome an industrial effort to automate as much as possible the process of modeling features and to jointly create a library of feature specifications.

6. Acknowledgements

The authors thank Gerard Holzmann at AT&T Bell Labs for many opportunities of discussions regarding Promela and SPIN. They also like to acknowledge his insights on how to enhance Promela to assert the receiving of a message from a channel. Thanks also to Ralph Blumenthal, E. Jane Cameron, Linda Hampton, Gary Herman, H. Paul Lin, Monagu Muralidharan, and R. C. Sekar for providing comments to improve this paper.

References

1. T. F. Bowen, F. S. Dworak, C. H. Chow, N. D. Griffeth, G. E. Herman, and Y.-J. Lin. The Feature Interaction Problem in Telecommunication Systems. In *Proceedings of the 7th International Conference on Software Engineering for Telecommunication Switching Systems*, pp.59-62, July 1989.

2. E. J. Cameron, N. Griffeth, Y.-J. Lin, M. E. Nilson, W. K. Schnre, and H. Velthuijsen. A Feature-Interaction Benchmark for IN and Beyond. *IEEE Communications Magazine*, pp.64-69, Vol. 31, No.3, March 1993.

3. Bellcore. *Advanced Intelligent Network (AIN) Release 0.2 Switching Systems Generic Requirements*. Generic Requirements GR-1298-CORE, Issue 1, November 1993.

4. Bellcore. *Denied Originating Service FSD 01-02-0301*. Technical Reference TR-TSY-000562, Issue 1, May 1990.

5. Bellcore. *Denied Termination FSD 01-02-0500*. Technical Reference TR-TSY-000563, Issue 1, May 1990.

6. Bellcore. *Call Waiting FSD 01-02-1201*. Technical Reference TR-TSY-000571, Issue 1, October 1989.

7. Bellcore. *Call Forwarding Subfeatures FSD 01-02-1450*. Technical Reference TR-TSY-000586, Issue 1, July 1989.

8. R. B. Blumenthal. *Feature Interaction Management - Report From the AIN Technical Collaboration*. Bellcore Technical Memorandum TM-NWT-023030, June 1, 1993.

9. E. J. Turner, Editor. *Using Formal Description Techniques - An Introduction to ESTELLE, LOTOS, and SDL*. Wiley, 1993.

10. G. Holzmann. *Design and Validation of Computer Protocols*. Prentice Hall, Englewood Cliffs, N.J., 1991.

11. Z. Manna and A. Pnueli. *The Temporal Logic of Reactive and Concurrent Systems, Volume 1 Specification*. Springer-Verlag, 1992.

12. Z. Manna and A. Pnueli. *Tools and Rules for the Practicing Verifier*. Technical Report Stan-CS-90-1321, Stanford University, 1990.

13. Z. Har'El and R. P. Kurshan. Software for Analytical Development of Communications Protocols. *AT&T Technical Journal*, pp.45-59, Vol. 69, No. 1, January 1990.

14. K. L. McMillan. *Symbolic Model Checking: An Approach to the State Explosion Problem*. PhD Thesis, Carnegie Mellon University, 1992.

15. J. R. Burch, E. M. Clarke, K. L. McMillan, and D. L. Dill. Symbolic Model Checking: 10^{20} States and Beyond. In *Proc. 5th Annual Symposium on Logic in Computer Science*, IEEE Computer Science Press, pp.428-439, June 1990.

16. P. Godefroid and G. Holzmann. On the Verification of Temporal Properties. *Protocol Specification, Testing, and Verification, XIII*, pp.109-124, North-Holland, 1993.

17. F. J. Lin and M. T. Liu. Protocol Validation for Large-Scale Applications. *IEEE Software Magazine*, pp.23-26, January 1992.

18. Bellcore. *Three Way Calling FSD 01-02-1301*. Technical Reference TR-TSY-000577, Issue 1, July 1989.

19. Bellcore. *Call Waiting Deluxe FSD 01-02-1215*. Technical Reference TR-NWT-000416, Issue 2, August 1993.

20. W. Bouma and H. Zuidweg. *Formal Analysis of Feature Interactions by Model Checking*. PTT Research Technical Report TI-PU-93-868, The Netherlands, 1993.

21. S. German. *Rapid Prototyping and Verification of Communication Services*. GTE Labs Technical Memorandum TM-0369-01-91-152, 1991.

22. R. Boumezbeur and L. Logrippo. Specifying Telephone Systems in LOTOS. *IEEE Communications Magazine*, pp.38-45, Vol.31, No.8, August 1993.

Appendix A. Originating Basic Call Model

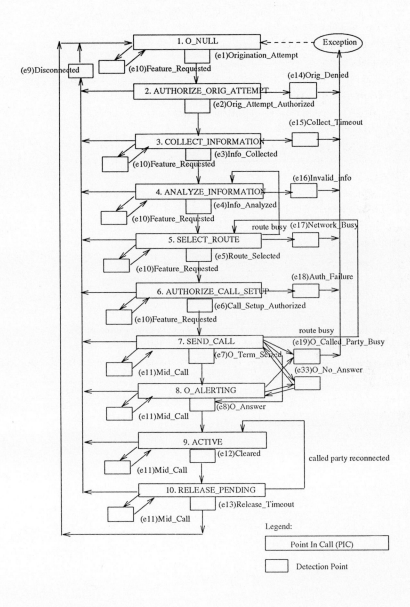

Appendix B. Terminating Basic Call Model

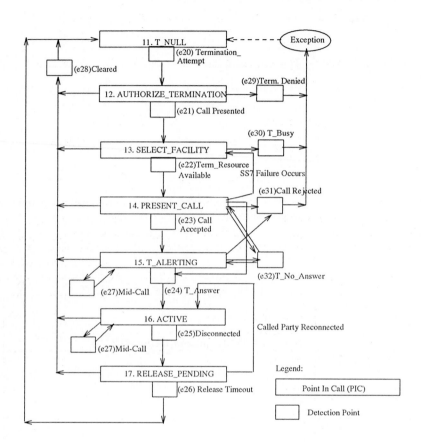

Appendix C. Two-Level Specifications for Terminating BFC

Part 1. Low Level Promela Specification for Terminating BFC

The model of terminating BFC consists of three processes and four channels as depicted below.

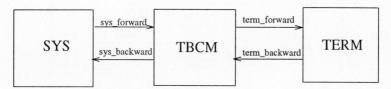

TBCM is the process modeling the terminating BCM, SYS modeling the switching system environment in which the BCM resides, TERM modeling the terminating subscriber's behavior. Channels sys_forward, sys_backward, term_forward, and term_backward are all unidirectional. The messages on each of them are listed in the table below.

Channel	Messages
sys_forward	call_setup(_) authorize(TRUE) authorize(FALSE) release(_) resource_ack(TRUE) resource_ack(FALSE)
sys_backward	auth_req resource_req release answer reanswer
term_forward	setup release ring
term_backward	alert(ACCEPTED) alert(SS7_FAILURE) alert(REJECTED) alert(ANSWER) alert(NO_ANSWER) answer(ANSWER) answer(NO_ANSWER) answer(REJECTED) feature_code(_) hangup(_) reanswer(_)

A typical scenario of the terminating call services is depicted below as an example.

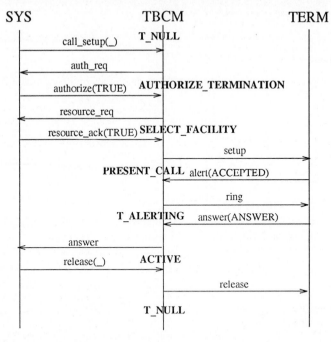

The complete specification of the terminating BFC in Promela is listed from the next page.

```
1  /**************************************/
2  /* For Supporting Both OBCM and TBCM */
3  /**************************************/
4
5  /* Boolean Values */
6  #define TRUE    1
7  #define FALSE   0
8
9  /* Auxiliary PICs */
10 #define END_DONE   99
11 #define HOLD       98
12 #define IDLE       0
13
14 /* Auxiliary DPs */
15 #define NULL       0
16
17 /* Dummy receiving variables */
18 int _;
19
20
21 /***********************/
22 /* For Supporting TBCM */
23 /***********************/
24
25 /* TBCM Message Parameter Values */
26 #define ACCEPTED      0
27 #define SS7_FAILURE   1
28 #define REJECTED      2
29 #define NO_ANSWER     3
30 #define ANSWER        4
31
32 /* TBCM PICs */
33 #define T_NULL                  11
34 #define AUTHORIZE_TERMINATION   12
35 #define SELECT_FACILITY         13
36 #define PRESENT_CALL            14
37 #define T_ALERTING              15
38 #define T_ACTIVE                16
39 #define T_RELEASE_PENDING       17
40
41 /* TBCM DPs */
42 #define TERMINATION_ATTEMPT       20
43 #define CALL_PRESENTED            21
44 #define TERM_RESOURCE_AVAILABLE   22
45 #define CALL_ACCEPTED             23
46 #define T_ANSWER                  24
47 #define T_DISCONNECTED            25
48 #define T_RELEASE_TIMEOUT         26
49 #define T_MID_CALL                27
50 #define T_CLEARED                 28
51 #define TERMINATION_DENIED        29
52 #define T_BUSY                    30
53 #define CALL_REJECTED             31
54 #define T_NO_ANSWER               32
55
56
57 /* OPTIONs */
58 #define UNSUC_SETUP       TRUE
59 #define T_EARLY_RELEASE   TRUE
60 #define SS7__FAILURE      TRUE
61 #define T_CALL_REJECTED   TRUE
62 #define NO__ANSWER        TRUE
63 #define T_IMMD_ANSWER     TRUE
64 #define T_BKWD_RELEASE    TRUE
65 #define T_REANSWER        TRUE
66 #define LINE_BUSY         TRUE
67
68 /* RCVD flags */
69 /* File RCVDsymbol to be included */
70
71 /* Queue Size */
72 #define QSIZE 2
73
74 /* Channels */
75 chan sys_forward   = [QSIZE] of {byte, bool};
76 chan sys_backward  = [QSIZE] of {byte};
77 chan term_forward  = [QSIZE] of {byte};
78 chan term_backward = [QSIZE] of {byte, byte};
79
80 /* initial state */
81 byte _state=T_NULL;
82
83 /* Busy Status */
84 bool _linebusy;
85
86 /* Message Types */
87 mtype = {
88 /* TBCM messages */
89     alert,
90     hangup,
91     resource_ack,
92     resource_req,
93     ring,
94 /* OBCM and TBCM common messages */
95     answer,
96     auth_req,
97     authorize,
98     call_setup,
99     connect,
100    feature_code,
101    reanswer,
102    release,
103    setup,
104    setup_cmp,
105    sub_busy,
106 };
107
108
109
110 byte sysid;
111 byte termid;
112 byte tbcmid;
113
114
115 proctype sys_model()
116 {
117 bool retry = FALSE;
118 bool auth;
119
120 idle:
121     if
122     :: _linebusy = FALSE
123 #if LINE_BUSY
124     :: _linebusy = TRUE
125 #endif
126     fi;
127     sys_forward!call_setup;
128
```

```
129 wait_to_authorize:
130     if
131     :: atomic
132     {
133         sys_backward?auth_req ->
134         if
135         :: auth = TRUE
136 #if UNSUC_SETUP
137         :: auth = FALSE
138 #endif
139         fi;
140
141         sys_forward!authorize(auth)};
142
143     if
144     :: auth -> goto wait_to_allocate
145     :: !auth -> goto end_done
146     fi
147 #if T_EARLY_RELEASE
148     :: sys_forward!release ->
149                 goto end_done
150 #endif
151     fi;
152
153 wait_to_allocate:
154     do
155     :: atomic
156     {
157         sys_backward?resource_req ->
158
159         if
160         :: !retry
161         :: retry ->
162         if
163         :: _linebusy = FALSE
164 #if LINE_BUSY
165         :: _linebusy = TRUE
166 #endif
167         fi
168         fi;
169
170         sys_forward!resource_ack(!_linebusy);
171         retry = TRUE;
172         _linebusy = TRUE
173     }
174     :: sys_backward?connect -> goto active
175     :: sys_backward?setup_cmp -> goto wait_to_allocate
176     :: sys_backward?sub_busy -> goto end_done
177 #if T_EARLY_RELEASE
178     :: sys_forward!release ->
179                 goto end_done
180 #endif
181     :: sys_backward?release -> goto end_done
182     od;
183
184 active:
185     if
186     :: sys_forward!release -> goto end_done
187     :: sys_backward?release -> goto wait_to_release
188     fi;
189
190 wait_to_release:
191     if
192     :: sys_forward!release -> goto end_done
193     :: sys_backward?connect -> goto active
194     :: goto end_done /* to create release timeout */

195     fi;
196
197 end_done:
198     do
199     :: sys_backward?auth_req
200     :: sys_backward?setup_cmp
201     :: sys_backward?resource_req
202     :: sys_backward?connect
203     :: sys_backward?release
204     od
205
206
207
208 }
209
210 proctype tbc_model()
211 {
212 bool auth;
213 bool all_right;
214 byte cond;
215 byte code;
216 do
217 :: _state == T_NULL ->
218     atomic
219     { sys_forward?call_setup(_) ->
220         /* e20: termination_attempt */
221         sys_backward!auth_req;
222         _state = AUTHORIZE_TERMINATION
223     }
224 :: _state == AUTHORIZE_TERMINATION ->
225     if
226     :: atomic
227     {
228         sys_forward?authorize(auth) ->
229         if
230         :: auth -> /* e21: call_presented */
231                 sys_backward!resource_req;
232                 _state = SELECT_FACILITY
233         :: !auth -> /* e29: termination_denied */
234                 _state = END_DONE
235         fi
236     }
237     :: atomic
238     {
239         sys_forward?release(_) -> /* e28: t_cleared */
240                 _state = END_DONE
241     }
242     fi
243
244 :: _state == SELECT_FACILITY ->
245     if
246     :: atomic
247     {
248         sys_forward?resource_ack(all_right) ->
249         if
250         :: all_right ->
251                 /* e22: term_resource_available */
252                 term_forward!setup;
253                 _state = PRESENT_CALL
254         :: !all_right ->
255                 /* e30: t_busy */
256                 sys_backward!sub_busy;
257                 _state = END_DONE
258         fi
259     }
260     :: atomic
```

```
261         {
262           sys_forward?release(_) -> /* e28: t_cleared */
263                    _state = END_DONE
264         }
265       fi
266
267 :: _state == PRESENT_CALL ->
268     if
269     :: atomic
270        {
271          term_backward?alert(cond) ->
272          if
273          :: cond == ACCEPTED    ->
274                       /* e23: call_accepted */
275                       term_forward!ring;
276                       sys_backward!setup_cmp;
277                       _state = T_ALERTING
278          :: cond == SS7_FAILURE ->
279                       /* ss7 failure occurs */
280                       sys_backward!resource_req;
281                       _state = SELECT_FACILITY
282          :: cond == REJECTED    ->
283                       /* e31: call_rejected */
284                       sys_backward!sub_busy;
285                       _state = END_DONE
286          :: cond == NO_ANSWER   ->
287                       /* e32: t_no_answer */
288                       _state = PRESENT_CALL
289
290          :: cond == ANSWER      ->
291                       /* e24: t_answer */
292                       sys_backward!connect;
293                       _state = T_ACTIVE
294          fi
295        }
296     :: atomic
297        {
298          sys_forward?release(_) -> /* e28: t_cleared */
299                    term_forward!release;
300                    _state = END_DONE
301        }
302     fi
303
304 :: _state == T_ALERTING ->
305     if
306     :: atomic
307        {
308          term_backward?answer(cond) ->
309          if
310          :: cond == REJECTED    ->
311                       /* e31: call_rejected */
312                       sys_backward!release;
313                       _state = END_DONE
314
315          :: cond == NO_ANSWER   ->
316                       /* e32: t_no_answer */
317                       _state = T_ALERTING
318
319
320          :: cond == ANSWER      ->
321                       /* e24: t_answer */
322                       sys_backward!connect;
323                       _state = T_ACTIVE
324          fi
325        }
326     :: atomic
327        {
328          sys_forward?release(_) ->
329                    /* e28: t_cleared */
330                    _state = END_DONE
331        }
332     :: atomic
333        {
334          term_backward?feature_code(code) ->
335                    /* e27: t_mid_call */
336                    _state = T_ALERTING
337        }
338     fi
339
340 :: _state == T_ACTIVE ->
341     if
342     :: term_backward?hangup(_) ->
343                    /* e25: t_disconnected */
344                    sys_backward!release;
345                    _state = T_RELEASE_PENDING
346     :: atomic
347        {
348          sys_forward?release(_) -> /* e28: t_cleared */
349                    term_forward!release;
350                    _state = END_DONE
351        }
352     :: atomic
353        {
354          term_backward?feature_code(code) ->
355                    /* e27: t_mid_call */
356                    _state = T_ACTIVE
357        }
358     fi
359
360 :: _state == T_RELEASE_PENDING ->
361     if
362     :: atomic
363        { sys_forward?release(_) ->
364                    /* e28: t_cleared */
365                    term_forward!release;
366                    _state = END_DONE
367        }
368     :: term_backward?reanswer(_) ->
369                    /* call party reconnected */
370                    sys_backward!connect;
371                    _state = T_ACTIVE
372     :: atomic
373        {
374          timeout ->
375                    /* e26: t_release_timeout */
376                    _state = END_DONE
377        }
378     fi
379
380 :: atomic
381    {
382      _state == END_DONE ->
383        _linebusy = FALSE;
384        break
385    }
386 od;
387
388 end_done:
389 do
390 :: term_backward?alert(cond)
391 :: term_backward?hangup(_)
392 :: term_backward?answer(cond)
```

```
393 :: term_backward?reanswer(_)
394 :: term_backward?feature_code(_)
395 od
396
397 }
398
399
400 proctype term_user()
401 {
402 byte cond;
403 byte count = 0;
404 end_idle:
405     do
406     :: atomic
407        { term_forward?setup ->
408            if
409            :: cond = ACCEPTED
410 #if SS7_FAILURE
411            :: cond = SS7_FAILURE
412 #endif
413 #if T_CALL_REJECTED
414            :: cond = REJECTED
415 #endif
416 #if NO_ANSWER
417            :: cond = NO_ANSWER
418 #endif
419 #if T_IMMD_ANSWER
420            :: cond = ANSWER
421 #endif
422            fi;
423
424            term_backward!alert(cond)};
425
426        if
427        :: cond == ACCEPTED -> goto wait_for_ring
428        :: cond == SS7_FAILURE -> goto end_idle
429        :: cond == REJECTED -> goto end_done
430        :: cond == NO_ANSWER -> goto end_done
431        :: cond == ANSWER -> goto talking
432        fi
433     od;
434
435 wait_for_ring:
436     do
437     :: atomic
438        { term_forward?ring ->
439            if
440            :: cond = ANSWER
441 #if T_CALL_REJECTED
442            :: cond = REJECTED
443 #endif
444 #if NO_ANSWER
445            :: cond = NO_ANSWER
446 #endif
447            fi;
448
449            term_backward!answer(cond)};
450
451        if
452        :: cond == REJECTED -> goto end_done
453        :: cond == NO_ANSWER -> goto end_done
454        :: cond == ANSWER -> goto talking
455        fi
456
457     :: term_forward?release -> goto end_done
458     od;
459
460 talking:
461     do
462 #if T_BKWD_RELEASE
463     :: term_backward!hangup ->
464                goto end_done
465 #endif
466 #if T_REANSWER
467     :: count < 2 ->      count = count + 1;
468                term_backward!hangup;
469                term_backward!reanswer;
470                goto talking
471 #endif
472     :: term_forward?release ->
473                goto end_done
474     od;
475
476 end_done:
477     if
478     :: term_forward?release
479     fi
480
481 }
482
483 init
484 {
485     sysid=run sys_model();
486     tbcmid=run tbc_model();
487     termid=run term_user()
488
489
490 }
491
```

Part 2. High Level Temporal Logic Specification for Terminating BFC

The table below is automatically generated from the temporal logic assertions database. The first column of the table identifies the option under which the terminating BFC's properties would exhibit. If the entry is a blank, the properties are considered standard. The second column of the table gives the informal English descriptions of the properties. The third column gives the corresponding temporal logic assertions. Each temporal logic assertion is prefixed with three flags of verification status in brackets, respectively indicating the success (V) or the failure (X) of (1) existence verification, (2) consistency verification, and (3) scenario generation. All the assertions below have passed the verification procedure. So they are all marked with [VVV].

Note that the primitives such as RCVDringTF, RCVDalterTB, RCVDresource_ackSF, RCVDresource_reqSB in the propositional formulas are the enhancements we did on top of Promela. They are used to assert that a message has just been received from a channel. Each primitive of this kind can be broken into into three parts.

— The first part is always an RCVD.

— The second part is used to indicate the name of a message. It is always in low case letters and understore such as ring, alter, resource_ack, and resource_req.

— The third part is used to denote the name of a channel. It can be two or three capital letters. In the terminating BFC, TF and TB denote channels term_forward and term_backward, respectively. SF and SB denote channels sys_forward and sys_backward, respectively.

For example, a propositional primitive RCVDringTF asserts that a message "ring" has just been received from the channel "term_forward".

\multicolumn{3}{c}{**Temporal Logic Assertions for Terminating BFC**}
Options	**Text Descriptions**	**Temporal Logic Assertions**
LINE_BUSY	— If the line is in an idle state when a terminating call is attempted, it should be placed in a busy state in preparation for completing the call.	— [VVV] (!_linebusy && sys_forward?[call_setup]) FOLLOWED BY (sys_forward?[resource_ack,TRUE]) IMPLIES EVENTUALLY (_linebusy)
	— If the line is in the busy state, the calling facility should be connected to busy tone.	— [VVV] (_linebusy && sys_forward?[call_setup]) FOLLOWED BY (sys_forward?[resource_ack,FALSE]) WITHOUT PRECEDING (sys_forward?[release]) IMPLIES EVENTUALLY (sys_backward?[sub_busy])
	— Ringing should be applied to the terminating line if it is found to be idle. — Audible ring tone should be provided to the calling facility while the called party is receiving ringing.	— [VVV] (!_linebusy && sys_forward?[call_setup]) FOLLOWED BY (RCVDresource_ackSF && tbc_model[tbcmid].all_right == TRUE) FOLLOWED BY (_state != END_DONE && RCVDalertTB && tbc_model[tbcmid].cond == ACCEPTED) IMPLIES EVENTUALLY (RCVDringTF)
		— [VVV] (!_linebusy && sys_forward?[call_setup]) FOLLOWED BY (RCVDresource_ackSF && tbc_model[tbcmid].all_right == TRUE) FOLLOWED BY (_state != END_DONE && RCVDalertTB && tbc_model[tbcmid].cond == ACCEPTED) IMPLIES EVENTUALLY (RCVDsetup_cmpSB)
T_EARLY_RELEASE	— The transmission path between the called line and the calling facility should be established as quickly as possible following initial detection of off-hook from the called line.	— [VVV] (!_linebusy && sys_forward?[call_setup]) FOLLOWED BY (_state != END_DONE && RCVDanswerTB && tbc_model[tbcmid].cond == ANSWER) IMPLIES EVENTUALLY (_state == T_ACTIVE)
	— When a disconnect signal is detected on the originating facility following the recognition of a seizure but before answer is received, further processing of the call should be stopped and the calling facility should be returned to the idle state.	— [VVV] (sys_forward?[release] && _state > T_NULL && _state < T_ACTIVE) FOLLOWED BY (!sys_forward?[release]) IMPLIES FOREVER (_state != T_ACTIVE) [VVV] (sys_forward?[release] && _state > T_NULL && _state < T_ACTIVE) FOLLOWED BY (!sys_forward?[release]) IMPLIES EVENTUALLY (_state == END_DONE)
	— When a disconnect signal is detected on the originating facility after the transmission path is established, the connection should be released and both the calling facility and the called line should return to the idle state.	— [VVV] (sys_forward?[release] && _state >= T_ACTIVE) IMPLIES EVENTUALLY (_state == END_DONE)

	Temporal Logic Assertions for Terminating BFC	
Options	**Text Descriptions**	**Temporal Logic Assertions**
T_REANSWER	— When a hangup is detected from the called line, the callee should be allowed to pick up the phone within the standard disconnect timing and re-establish the connection.	— [VVV] (_state == T_ACTIVE && RCVDhangupTB) FOLLOWED BY (_state == T_RELEASE_PENDING && RCVDreanswerTB) IMPLIES EVENTUALLY (_state == T_ACTIVE)
T_BWKD_RELEASE	— If a hangup is detected from the called line, the connection will be released after an interval that exceeds the standard disconnect timing.	— [VVV] (_state == T_ACTIVE && term_backward?[hangup]) FOLLOWED BY (RCVDtimeout) IMPLIES FOREVER (_state != T_ACTIVE)
		— [VVV] (_state == T_ACTIVE && term_backward?[hangup]) FOLLOWED BY (RCVDtimeout) IMPLIES EVENTUALLY (_state == END_DONE)
UNSUC_SETUP	— The call should be stopped if it is not authorized during the call setup.	— [VVV] (!_linebusy && sys_forward?[call_setup]) FOLLOWED BY (sys_forward?[authorize,FALSE]) IMPLIES EVENTUALLY (_state == END_DONE)
UNSUC_SETUP	— The call should be stopped if there is no terminating facility available during call setup.	— [VVV] (!_linebusy && sys_forward?[call_setup]) FOLLOWED BY (sys_forward?[resource_ack,FALSE]) IMPLIES EVENTUALLY (_state == END_DONE)
SS7_FAILURE	— The call should be retried if a SS7 failure occurs after outpulsing.	— [VVV] (_state == PRESENT_CALL) FOLLOWED BY (_state != END_DONE && RCVDalertTB && tbc_model[tbcmid].cond == SS7_FAILURE) IMPLIES EVENTUALLY (RCVDresource_reqSB)
T_IMMD_ANSWER	— Certain terminating facilities can decide to answer, not to answer, or to reject the call after being setup but before being rung. Upon being answered, the connection should be established immediately.	— [VVV] (_state == PRESENT_CALL) FOLLOWED BY (_state != END_DONE && RCVDalertTB && tbc_model[tbcmid].cond == ANSWER) IMPLIES EVENTUALLY (_state == T_ACTIVE)
NO_ANSWER	— Upon no answer, the call should never be connected through.	— [VVV] (_state == PRESENT_CALL) FOLLOWED BY (term_backward?[alert,NO_ANSWER]) IMPLIES FOREVER (_state != T_ACTIVE)
T_CALL_REJECTED	— Upon being rejected, the call should be torn down.	— [VVV] (_state == PRESENT_CALL) FOLLOWED BY (term_backward?[alert,REJECTED]) IMPLIES FOREVER (_state != T_ACTIVE)
T_CALL_REJECTED	— The call should be stopped if it is rejected by the terminating facility during ringing.	— [VVV] (_state == T_ALERTING) FOLLOWED BY (term_backward?[answer,REJECTED]) IMPLIES FOREVER (_state != T_ACTIVE)
NO_ANSWER	— The call should never be connected through if no answer is provided by the terminating facility during ringing.	— [VVV] (_state == T_ALERTING) FOLLOWED BY (term_backward?[answer,NO_ANSWER]) IMPLIES FOREVER (_state != T_ACTIVE)

Formalisation of a User View of Network and Services for Feature Interaction Detection

Pierre Combes, Simon Pickin

CNET PAA/TSA/TLR

38 rue du général Leclerc, 92331 Issy-les-moulineaux, FRANCE

email: pierre.combes@issy.cnet.fr   simon.pickin@issy.cnet.fr *

Abstract. This paper proposes a method for detecting feature interactions by verifying that service feature requirements, specified in a formal property language (temporal logic), are satisfied on a formal specification language model (in SDL) of the network and the service features concerned. We develop an abstract model, representing a user (external) view, of the network and the service features, we give examples of IN feature requirements together with their expression in a temporal logic language and we discuss the choice of an efficient property language.

1 Introduction

1.1 The Feature Interaction Problem

The feature interaction problem is now well known in the telecommunications world as a major obstacle to the rapid deployment of new telecommunications services and features [21, 2, 1, 20]. Indeed, it is expected to give rise to new major difficulties in understanding and mastering how service features cooperate and interact, how they share resources and what the resulting behaviour of the network is. Each new feature may interact with many existing features, causing customer annoyance or total system breakdown, and we cannot assess a priori that [20]:

- for the customer (subscriber or end-user), the features shall behave properly as regards information received from the network (expected to be correct and user-friendly), information sent to the network (expected to be taken into account), waiting time (expected to be short) and results (expected to match each feature's requirement);

- for the service provider, the features shall remain available, reliable and in accordance with user requirements, data shall remain consistent, and the call records shall permit the subscriber to be charged for the use of these features;

- for the network operator, the services shall not create any deadlock or abnormal resource consumption in the network (and charging should be consistent).

---

*The work in this paper has been carried out in the RACE project SCORE, ref. 2017. It reflects the view of the authors.

1.1.1 Definition of interaction

To give a precise and complete definition of feature interaction is a perilous task in view of the huge variety of problems which could be classified under this heading. As a working definition, we [12, 20, 11] use a reformulation of the definition of [7], stating that there is interaction:

- when a service feature inhibits or subverts the expected behaviour of another service feature (or of another instance of the same service feature).

- when the joint accurate execution of two service features provokes a supplementary phenomenon which cannot happen during the processing of each of the service features considered separetely.

This interaction definition focuses on the manifestation of interactions. In fact, every new feature introduced in the network interacts with existing features but we reserve the term interaction for the case where unexpected behaviour occurs.

1.2 Service Creation in SCORE

The main goal of SCORE is to define and prototype a set of methods and tools to be incorporated into service creation environments, the overall objectives being:

- Define a model of the service creation process capable of handling its full complexity and adapted to developing services for an open environment.

- Based on this process model, define and prototype a service creation environment consisting of a set of methods and tools.

- Define a framework of appropriate interfaces for service creation methods and tools, notably by defining a component model.

- Validate, through a realistic application, the service creation process model, the methods and tools and their interfaces.

- Initiate and support the exploitation of the project results, and contribute to standards in the domain of service creation.

In the development of methods and tools for service creation, SCORE pays special attention to the feature interaction problem, focusing on detection of interactions during the analysis, specification and design activities of the service creation process.

1.2.1 Focusing on Logical Interaction

During the service creation process, a feature will be described at several different levels of abstraction, from a high level view to implementation code. Causes of interactions can be associated to these different levels. Adapted from [7], we call interactions occuring at the level of the abstract specification, *logical interactions*, those occuring when the feature specifications are mapped onto a network architecture, *network interactions* and those occuring when the feature software is mapped onto an execution environment, *implementation interactions*.

Interactions should be detected as early as possible, particularly logical interactions which will otherwise propagate through all the service creation activities. Therefore, though SCORE also considers design interactions [12], this paper focuses on logical interactions.

1.3 The need for formal methods and tools

Traditionally, interactions were detected and resolved on a feature by feature basis by experts knowledgeable on all existing features. However, in the future this approach will not be practically applicable, since the problem grows at least exponentially with the number of features. Features themselves are becoming more and more complex and new service features are continually being added to improve or extend them. Moreover, in the near future, independent parties will be involved in the specification, design, implementation and deployment of services and features which will also be deployed in the network by different operating companies and their associated vendors. This calls for more precise specifications and improved methods for integrating features which have been independently conceived and specified into a unified design.

Formal specifications facilitate exchanges between different actors and can be used for more coherent standardisation. In addition, formalisation allows a better understanding of the behaviour of features and of requirements on this behaviour. Methods for avoiding or resolving interactions (during feature invocation by some negotiation agent model [17], or during feature management [20]), are more easily elaborated if these interactions have been previously analysed. Moreover, we cannot expect to solve all interaction problems while satisfying the requirements of all the different actors involved: *perfection is not of this world* [20]. Only a precise and formal specification of the features in the network can help to give a common and acceptable understanding of the resulting behaviours, enabling a coherent choice between different solutions to be proposed to the different actors involved[1].

1.3.1 The SDL Specification Language

Though other formal specification languages such as LOTOS could be used for such a method [5], the desire to conform to ITU-T standards with wide acceptence among telecommunication operators, as well as the existence of commercial tools, dictated the choice of SDL inside the SCORE project.

The SDL language is based on the notion of Extended Finite State Machine. An SDL specification is composed of parallel processes communicating by FIFO queues, using an abstract data type approach to data specification.

2 Overview of the Detection Method

2.1 General Principle

To detect occurence of feature interactions, we develop a method based on expression of feature requirements as properties in a property language and on detection of interactions as the system's inability to satisfy these properties [12, 11, 5].

More precisely, let $F_1, F_2, \ldots$ resp. N be user-view specifications of features, resp. the underlying network, in some constructive specification language (in our case SDL). Let $N \oplus F_i$ denote the specification (in this same language) of the (composite) network obtained by adding service feature F_i to the basic network N. Recursively, let $\mathcal{N} \oplus F_i$ be the network obtained by adding feature F_i to (possibly composite) network $\mathcal{N}$. Let

---

[1] we do not use the concept of wanted or unwanted interactions since we cannot say whether an interaction is wanted or unwanted before having detected and analysed it, and since a behaviour acceptable to one actor could be unacceptable to another

$P_1, P_2, \ldots$ be formulae expressing feature requirements in a suitable property language and let $\mathcal{N} \models P$ denote that the network specification $\mathcal{N}$ satisfies, i.e. is a model of, formula P.

To detect an interaction between features, one first formalises the expected properties, P_i of each feature F_i. We then say that there is interaction between features $F_1, \ldots, F_n$ if:

$$N \oplus F_i \models P_i \quad, \quad 1 \leq i \leq n$$

but

$$N \oplus F_1 \oplus \ldots \oplus F_n \not\models P_1 \wedge \ldots \wedge P_2$$

Note that this method depends heavily on careful and, if possible, complete formalisation of expectations. We will usually consider the case where $n = 2$ under the assumption that most interactions involve only two service features and can therefore be detected by considering features pairwise.

2.2 User-view abstract model

The abstract network model represents an external view (i.e. as seen by subscribers or end-users) of an IN network based largely on notions taken from the ITU-T concepts of the Service Plane (though recommendations here are very weak) and the Global Functional Plane (basic call, detection point,...) as defined in [9].

The Basic Call Process (BCP) is modelled as an SDL process, each call being an instantiation of this process. Feature behaviours in the network are modelled by concurrent SDL processes which communicate with the BCP by Points of Initiation (POI) and Points of Return (POR) [9]. Currently, in this approach, the features do not communicate between themselves. Invocation of an IN feature is modelled as an output action, occuring at a Detection Point in the BCP, to the feature process. At the end of the execution of the feature process, an output action to the BCP occurs, for which there will be a corresponding input action occuring at the start of the appropriate Point In Call of the BCP. Specification of the communication mechanisms is elaborated in section 4.3.

2.3 Properties of Service Features

To be able to verify the correctness of the system behaviour we have to express formally the feature requirements and general network properties. Thus we need a formal property language allowing expression not only of general properties, such as invariants and deadlocks, but also specific properties. Such properties often concern the temporal ordering of events, e.g. safety, liveness or precedence properties. A common choice for such a language is *Modal or Temporal Logic* [19, 25, 10, 13, 14, 24].

To each feature, whose behaviour is modelled as an SDL process, we associate formal properties expressing basic requirements on its behaviour. Though these formal properties are an abstraction of the user's perception of feature requirements, the fact that they are applied to an SDL specification means that the specification of properties depends on the semantics of SDL (see section 3.1). In addition to expressing subscriber requirements, the formal properties associated to a feature may also express requirements of the Service Provider, the Network Provider or those of other users (see example of code protection in section 4.2).

2.4 Verification Tools

Verification of the formal properties of the features and the network is performed by the SDL tool GEODE [15, 3] using animation and state-space exploration techniques.

- Animation tools can be used to symbolically execute service descriptions. This technique is widely used today, because it allows validation of large systems. However, it provides only a partial proof about the behaviour of these systems.

- State-Space exploration techniques check during the construction of the graph that some properties expressed in a formal way are verified. Though, based on exhaustive simulation, some violations of properties may be detected before the complete state space has been explored.

- Also of interest is Model Checking. In this technique, first the graph of all reachable states of the system is constructed, then the satisfaction of formal properties on it is automatically checked. The construction of the entire graph restricts the possible use of this technique to systems with finite behaviours and the practical use to small systems, due to the so-called state explosion [16] (the exponential increase in the size of the state graph). This technique was experimented in a first phase with the tool XESAR [22, 11].

3 Expression of Service Properties

3.1 The Temporal Logic Framework

3.1.1 The Linear Temporal Logic Operators

The temporal logic we use for the formal expression of IN feature requirements is a linear time temporal logic [19, 18] since, we have not found any real need for the existential operator on paths (used to express potentiality) which distinguishes branching time temporal logic [2]. However, this choice is made for the relatively simple models and verification procedures currently used and is therefore not definitive; for a more elaborate discussion about the different temporal logics see section 3.3.

A temporal logic language is defined over infinite sequences of states, representing execution states of the specification. The operators of this logic with their intended (and informal) meaning are as follows, where as in [18], we define these operators to be irreflexive [3]:

- Always Operator: **G**(a) : a is true at all states after the reference state.

- Eventually Operator: **F**(a) : a is true at at least one state after the reference state.

- Next Operator: **NEXT**(a) : a is true at the state just after the reference state.

- Until Operator: a **UNTIL** b : a is true at all following states up to a state at which b is true.

---

[2]The following formal properties can be transposed to branching time temporal logic (CTL*), by substitution of the initial **G** operator by the branching time operator **AG**.

[3]roughly speaking, the present is not included in the future

Starting from the definition of the **UNTIL** binary operator, we can introduce some derived operators :

- a **ATNEXT** b : a holds at the next future state where b holds:

$$a \text{ ATNEXT } b == (\text{ not } b \text{ UNTIL } (a \text{ and } b)) \text{ or } (\text{G not } b)$$

- a **BEFORE** b : If b holds in some future state, a holds in an earlier state:

$$a \text{ BEFORE } b == \text{not } ((\text{not } a) \text{ UNTIL } b)$$

- a **ATNEXT**^{n} b : a holds at the nth occurence of b:

$$a \text{ ATNEXT}^2 b == (a \text{ ATNEXT } b) \text{ ATNEXT } b$$

3.1.2 State Formulas

General temporal properties are constructed by applying the temporal operators to the formulas of the state language, these being logical combinations of state predicates. It is not the scope of this paper to completely define a formal semantics of an SDL property language (see, for example [24]). However, to allow the following examples of property expression to be understood, we need a definition of basic SDL objects and state predicates [24, 6].

Data predicates : boolean expressions on the specification data, including description of SDL channels and FIFO queues.

Control predicates : e.g. for a SDL state or label, $\mathbf{at}(Pi, Sj)$ is true for the global states q in which the process Pi is on the location Sj.

event predicates These predicates concern the SDL actions, in particular for communication actions like SDL INPUT-OUTPUT: $\mathbf{ex}(P_i, a)$ is true on the global states q from which the action a of the process P_i is executed in the next execution step [4]. In the examples we simplify $\mathbf{ex}(\text{call}_i, a)$ by $\mathbf{ex}(a_i)$.

Some SDL objects

Each call is an instance of the basic call process, consequently has an identifier and the following parameters: calling number, digit number dialled, called number, destination number. We write **call$_i$(a,b,c,d)** where

$$\text{call}_i(a,b,c,d) == (\text{calling}(\text{call}_i) = a) \text{ AND } (\text{digit-dialled}(\text{call}_i) = b) \quad (1)$$
$$\text{AND } (\text{called}(\text{call}_i) = c) \text{ AND } (\text{dest}(\text{call}_i) = d)$$

We also use macro-notations for subclauses of the above conjunction, e.g.

$$\text{call}_i(a,,,d) == (\text{calling}(\text{call}_i) = a) \text{ AND } (\text{dest}(\text{call}_i) = d) \quad (2)$$

---

[4]In the temporal μ-calculus [24, 25], it corresponds to the expression using the relativised next operator $\mathbf{X}_a(\text{tt})$

Communication signals user to network: off-hook(x), senddigits(x,y) (digit/s y is/are sent from subscriber no. x), busy(x), no-answer(x), on-hook(x). Network to user: connect(x,y) (call attempt made from subscriber number x to destination number y), end-of-call(x,y),

Basic Call states : start state, conversation state, ... (e.g. $at(call_i, conv)$). To allow more readable formulas we introduce the macro-notation:
at-conv$_i$(x,y) == at (call$_i$(x,,,y), conv)
stating that subscribers x and y are in conversation.

3.1.3 A useful macro-operator

For many user properties we want to express that an event should be followed immediately by another event, i.e. we need a stronger operator than F. In the simple models we are concerned with, we can do this using the notion of the list of states of a Basic Call:
l-call-states == (idle, collect-info, analyse-info, select-route, ... conv)
We define the macro:
SOON$_i$(action) == (action) **BEFORE** at(call$_i$,l-call-states)
expressing that the action happens before the BCP arrives in its next state, one of the states in *l-call-states* (i.e. before the next trigger detection point). This operator is especially useful for service features whose POI and the POR are between the same basic states of a call. Of course, its usefulness presupposes the property 15 that all calls must eventually make progress, this being one of the overall properties which must also specified and verified.

3.2 Formal Feature Properties

Call Waiting this feature enables the subscriber to be notified of an incoming call when his/her line is engaged. (In reality this feature is more complex; the user could, for example, admit the third party to the conversation, etc. Here we define its most basic requirement).
We introduce the state predicate *active-CW(x)*, true if the CW feature is activated for the subscriber x, and the signal *signal-CW* indicating to subscriber x that an incoming call has arrived. Then, for $i \neq j$, we require:

$$\textbf{G} ((\text{at-conv}_i(x,z) \textbf{ and } \text{active-CW}(x) \textbf{ and } \text{ex}(\text{connect}_j(y,x))) \Rightarrow \quad (3)$$
$$\textbf{SOON}_j \text{ ex}(\text{signal-CW}(x))$$

Completion of Calls to Busy Subscribers allows the calling user (the subscriber) encountering a busy destination to be informed when the busy destination becomes free (property 4).

Call Completion on No Reply allows the calling user (the subscriber) encountering a no answer in the destination line to be informed when the destination line is next used (on its release). The subscriber is informed when the destination line is released (property 5). In fact, requirements 4 and 5 are very similar:

$$\textbf{G} ((\text{call}_i(x,,,y) \textbf{ and } \text{ex}(\text{busy}_i(y)) \textbf{ and } \text{active-ccbs}(x)) \Rightarrow \quad (4)$$

$$(\textbf{SOON}_j \ (\textbf{ex}(\text{ccbs-recall}(x))) \ \textbf{ATNEXT} \ \textbf{ex}(\text{on-hook}_j(y))) \ )$$
$$\textbf{G} \ (\ (\text{call}_i(x,,,y) \ \textbf{and} \ \textbf{ex}(\text{no-answer}_i(y)) \ \textbf{and} \ \text{active-ccnr}(x)) \Rightarrow \quad (5)$$
$$(\textbf{SOON}_j \ (\textbf{ex}(\text{ccnr-recall}(x))) \ \textbf{ATNEXT} \ \textbf{ex} \ (\text{on-hook}_j(y))) \ )$$

Note that the index of the operator **SOON** is the same as that of the *on-hook* event. The events *ccbs-recall(x)* and *ccnr-recall(x)* are the events which trigger the respective features and which take place between the *on-hook* and the final state of the relevent BCP.

Temporary deactivation of the feature (CW) allows a subscriber to the CW feature to deactivate this feature during the current call.

$$\textbf{G} \ ((\text{at-conv}_i(x,y) \ \textbf{and} \ \textbf{ex}(\text{desact-temp-cw}(x)) \ ) \ \Rightarrow \quad (6)$$
$$(\textbf{not}(\textbf{ex}(\text{signal-CW}_j(x))) \ \textbf{UNTIL} \ \textbf{ex}(\text{end-of-call}_i(x,y)))))$$

Automatic Call Back allows the subscriber, by pressing the ACB code, to set up a call to the calling party of the last call connected to his/her line.

$$\textbf{G} \ (\ (\text{active}_{acb}(x) \ \textbf{and} \ \textbf{ex}(\text{connect}_i(y,x)) \ \textbf{and} \quad (7)$$
$$(\ \textbf{not} \ \textbf{ex}(\text{connect}_j(z,x))) \ \textbf{UNTIL} \ \textbf{ex}(\text{senddigits}_v(x,\text{code}_{acb}))) \Rightarrow$$
$$(\textbf{SOON}_v \ \text{called}(\text{call}_v) = y \ \textbf{ATNEXT} \ (\textbf{ex}(\text{senddigits}_v(x,\text{code}_{acb}))) \ )$$

Where $i \neq j$. It states that if a call attempt is made from y to x while the feature is activated (first line) and no other call attempts arrive at x before the next use of the ACB code (second line) then on this next use of the ACB code at x, which invokes a call instance identified by v, before instance v reaches its next state, the called party of v is set to y (third line).

Terminating Call Screening allows the subscriber to specify that incoming calls be rejected or allowed, according to a screening list (property 8).

Originating Call Screening allows the subscriber to specify that he/she will never be connected as a calling party in a call for which the destination number is on a screening list (property 9).

$$\textbf{G} \ \textbf{not} \ (\ (y \ \text{in} \ \text{TCS}(x)) \ \textbf{and} \ \textbf{ex}(\text{connect}_i(y,x))) \quad (8)$$
$$\textbf{G} \ \textbf{not} \ (\ (y \ \text{in} \ \text{OCS}(x)) \ \textbf{and} \ \textbf{ex}(\text{connect}_i(x,y))) \quad (9)$$

ABD service The subscribers can associate short digit codes with directory numbers of their choice. Composing this digit code is equivalent to dialling the full directory number associated to it (property 10).

Simple Speed Calling allows the subscriber to make an automatic call to a preselected number without numbering (property 11).

$$\textbf{G} \ (\ \textbf{ex}(\text{senddigits}_i(x,\text{ABD}_n)) \Rightarrow \quad (10)$$
$$\textbf{SOON}_i \ (\text{call}_i(x,\text{ABD}_n,\text{TR}(\text{ABD}_n),)))$$
$$\textbf{G} \ (\ \textbf{ex}(\text{senddigits}_i(x,0)) \Rightarrow \quad (11)$$
$$\textbf{SOON}_i \ (\text{call}_i(x,0,\text{predef}(x),)))$$

Call Forwarding Unconditional We can find two different definitions of this feature [12]:
If a user has CFU set to a certain number, then all calls to the user's number will be connected to the forward number (property 12).

$$\mathbf{G}\ (\text{call}_i(x,,y,)\ \mathbf{and}\ \text{active}_{cfu}(y)\ \mathbf{and}\ CF_y(z) \Rightarrow \qquad (12)$$
$$\mathbf{SOON}_i\ \text{call}_i(x,,y,z))$$

If a user has CFU set to a certain number, then during all call set ups to the user's number a translation of the user's number to the forwarded number will occur (property 13).

$$\mathbf{G}\ (\ \text{call}_i(x,,,y)\ \mathbf{and}\ \text{active}_{cfu}(y)\ \mathbf{and}\ CF_y(z) \Rightarrow \qquad (13)$$
$$\mathbf{SOON}_i\ \text{call}_i(x,,,z))$$

We remark that property 12 does not allow sequencing of multiple forwards while property 13 can imply an interaction between two instances of this feature. This interaction will be detected by violation of property 15.

Call Forwarding on Busy Line similar to the definition of Call Forwarding but this feature is invoked only if the destination line is busy.

$$\mathbf{G}\ (\text{call}_i(x,,,y)\ \mathbf{and}\ \mathbf{ex}(\text{busy}_i(y))\ \mathbf{and}\ \text{active}_{cful}(y)\ \mathbf{and}\ CFBL_y(z) \qquad (14)$$
$$\Rightarrow \mathbf{SOON}_i\ \text{call}_i(x,,,z))$$

Basic Call Property : No local deadlock during Call Teatment.

$$\mathbf{G}\ (\ \text{off-hook}_i(x)\ \Rightarrow \mathbf{F}\ (\text{at-conv}_i(x,)\ \text{or at-idle}_i\ ) \qquad (15)$$

3.3 Which temporal logic

Many classes of Temporal/Modal logic have been defined for temporal reasoning and verification tools. We can classify them in three main categories:

- Linear Time temporal Logic (LTL) [18, 19] where at each moment there is only one possible future.

- Full Branching Time Temporal Logic (CTL*)[14] where at each moment time may split into alternative course representing different possible futures. In this logic we define *state-operators* which hold at a certain state irrespective of the paths through that state, and *path-operators* which depend on a particular path (closed to linear operator). State operators are:
  Ap : The formula Ap is true at a state if p is true on all paths starting at that state.
  Ep : The formula Ep is true at a state if p is true on some path starting at that state.

- Branching Time Temporal Logic (CTL) [10, 22], subset of CTL* obtained by imposing restrictions on the order in which temporal operators may be placed next to each other: state-operators and path-operators should be applied alternately.

The choice of a suitable temporal logic should be based on the following requirements: expressivity, readibility and efficiency for tools.

expressivity : CTL* is the most expressive of these logics. CTL, being a restriction of CTL*, is less expressive and in particular cannot express fairness properties like GF. In LTL we cannot express properties about the notion of possibility, though, as already stated, it seems that such a concept is not needed for user-view feature requirements. This is since user properties tend to follow the scheme:
action => **eventually** *effet*   or   *action* => **always** *property*.
Properties such as *It is always be possible to go back to the initial state*, will be obtained by the definition of a possible user action realising this wish at certain stable states (l-stable-states) together with the properties:
G F (**at**(*l-stable-states*)) and
G (**at**(*l-stable-states*) **and** **ex**(*return-action*) ⇒ F **at**(*init-state*))

readibility : Readibility incorporates two sometimes opposed characteristics: the ease with which a certain requirement can be expressed in the language and the ease with which a formula can be seen to correctly represent a certain requirement. The restricted expressivity of CTL makes it a complex language to express properties (e.g. how to express property 8), a problem which is alleviated to some extent by the introduction of derived operators. The expressivity of CTL* means that it sometimes violates the second characteristic. A comparison between CTL* and LTL is more subjective,

efficiency for tools :

- CTL is the most efficient logic for techniques such as model checking (algorithms are linear in the size of formulas).
- LTL is the most efficient logic for development of proof checking techniques, and resolution methods have been developed for such languages [8][5]. In addition, an LTL property can easily be translated into an SDL automata verifying it on infinite sequences; such an approach will allow verification of some temporal logic properties with state-space exploration tools.

Finally, we do not intend to directly use the **NEXT** operator in properties, this operator being too related to the notion of a next step in the execution of an SDL specification. Finally, the use of past operators [19] could also be considered; in particular, properties like 8 could be more easily expressed with **ATLAST**, a past operator similar to the **ATNEXT** future operator.

4 Developments of the Model

4.1 An Evolutive Model

Formalisation depends on identification of the necessary concepts to be introduced and the relationships between them. From a first analysis [11] of IN features such as Call Forwarding, Terminating Call Screening, Follow-Me-Diversion, Terminating Code

---

[5] It will be probably of great benefit to investigate further the use of such techniques to complement our method. Indeed, many interactions could be detected by proof of inconsistencies between feature properties.

Protection, Automatic Call Back we identified necessary concepts such as Directory number, Physical line, Personal Identification Number, protection codes and several different "White or Black Lists"[6].

However, analysis of the feature interaction problem shows that many interactions are due to the fact that the introduction of a new feature can modify some basic network concepts, and consequently the definition of existing features[20]. For example, redirection features (Call Forwarding, UPT) introduce a distinction between the directory number and physical line. Interactions can then arise because some implicit assumptions taken during the definition of an existing feature become explicit (see [11] and interaction between CW and TCS features).

In the future, with the introduction of new and complex features in the network, the abstract network model will have to be evolutive in order to take into account this continual modification of the user view of the network. The use of object-oriented methods and tools (such as OMT [23]) will be of great benefit during the analysis and specification of the network together with new features. Note however that the feature properties will usually be more stable than the constructive SDL specifications.

4.2 Completeness of the model

The main objective of the POTS network was to connect users (due to this basic concept, taxation was only applied to completed calls), whereas in IN, various different types of information can be exchanged between users or between network and users, e.g. consider features with vocal interfaces. Since feature interaction can result in inconsistency or ambiguity in these information exchanges these new concepts need to be reflected in the formal model. In the case of the vocal interfaces for example, a formal semantics must be attached to the associated questions and answers. To illustrate this point, consider:

1. Interaction between Calling Number Delivery feature which delivers the directory number of the calling party to the destination party, and Unlisted Number feature which allows a subscriber to keep his number private.

2. Interaction between Call Forwarding and Code Protection features. The code protection feature allows a subscriber to protect his/her line by a user-defined key, i.e. PIN code, which callers are required to enter after the network has delivered a message asking for the correpondent's code. Suppose subscriber a forwards his calls to subscriber b who has the code protection feature active. When a user c calls a, his/her call will arrive at b's line whereupon he/she will be asked to give a protection code. However, c may not know this code, nor even the destination line of his/her call.

3. Associated to the Call Forwarding feature is the subscriber requirement: he/she does not want it to be known that his/her calls are forwarded (e.g. for burglar protection). This requirement can be violated by other features which deliver information about the destination. This information may be implicit as in the case of the change of subscriber code discussed above. It may also be explicit as in the case where the previous interacton is resolved by a message telling the calling party the number to which the code they must enter is associated.

---

[6] Notice that such a model will have many similarities with a "user view" of a distributed software system: physical entities (telephone physical line) shared by different users, login, password, access protection, data and resource sharing, etc.

These interactions raise an issue whose importance is likely to increase: we need to consider the notion of user's/subscriber's knowledge of the characteristics of his/her environment. One approach to this problem would be to treat this notion of knowledge with the use of modal knowledge operators and perhaps to combine them with temporal operators in a multimodal logic. Currently we adopt the more ad-hoc approach outlined below. This notion of knowledge adds new, possibly implicit, requirements on the behaviour of a feature. In the case of code protection the specification of the requirements should take into account that a major functionality of code protection is that this code be known to some users and unknown to others. If we specify the knowledge domain of each user, we can then express properties which specify that some information should not be given to a user. In the case of code protection, we define two properties, the first one for the subscriber of the feature who does not want anyone unaware of his/her code to reach him/her, the second for other users who, if they know the code, expect it to be accepted.

The SDL specification of the Code Protection feature includes the variable t, the subscriber no. which y believes to be the code owner, the function $know\text{-}code(x,t)$, indicating that subscriber x knows the code of subscriber t, the message $ask\text{-}code_i$ associated with the POI of the feature, and the message $ans\text{-}code(b)$ associated with the POR of the feature, the boolean variable b being false if the code protection feature detects that the calling party does not give the correct code. The two properties mentioned above, one for the subscriber to the feature, z, and one for the calling party, x (the second of these properties will not be satisfied by the specification unless $t = z$ whereas in fact in the specification $t = y$) are:

$$\text{G } (\ (\ \text{call}_i(x,,,z) \text{ and } \text{ex}(\text{ask-code}_i(z)) \text{ and not } know\text{-}code(x,z) \ ) \quad (16)$$
$$\Rightarrow \text{G not}(\text{ex}(\text{ans-code}_i(\text{true}))))$$

$$\text{G } (\ (\ \text{call}_i(x,,y,z) \text{ and } \text{ex}(\text{ask-code}_i(z)) \text{ and } know\text{-}code(x,t) \ ) \quad (17)$$
$$\Rightarrow \text{SOON}_i \ \text{ex}(\text{ans-code}_i(\text{true})))$$

To express properties about exchanges of information between network and users, we introduce the action $sendinfo(x,i)$ (info i to user no. x). We then use it to express properties for the Calling Number Delivery feature (property 18) and the Unlisted Number feature (property 19) (with the definition given below):

$$\text{G } (\ (\text{call}_i(x,,,z) \text{ and } \text{active}_{cnd}(z)) \Rightarrow \quad (18)$$
$$\text{SOON}_i \ \text{ex } (\text{sendinfo}(z,x)))$$

$$\text{G NOT } (\text{activ}_{un}(x) \text{ and } \text{ex}(\text{sendinfo}(y,x))) \quad (19)$$

To express the Call Forwarding requirement *nobody should know that my calls are fowarded*, we must make this notion more precise (within our current model). In other words, we need to formalise which information given by the network will violate this requirement (destination number, special tone, ...). With the destination number as the only violating information we have:

$$\text{G NOT } (\text{active-CF}(y) \text{ and } \text{call}_i(x,,y,z) \text{ and } (\text{sendinfo}(x,z) \text{ and } z \neq y)) \quad (20)$$

Now suppose we resolve the Call Forwarding and Code Protection interaction by giving the calling party the number to which the code they must enter is associated when asking for the code. This means that the feature specification now satisfies:

$$\text{G } (\ \text{call}_i(x,,,z) \text{ and } \text{ex}(\text{ask-code}_i(z)) \Rightarrow \quad (21)$$
$$\text{SOON}_i \ (\text{ex}(\text{sendinfo}(x,z)) \text{ and } (t=z) \ )$$

We can then be sure of satisfying the calling party's requirement on the Code Protection feature 17 in the general case since now $t = z$ even if $y \neq z$. However, an interaction can then be detected between the call forwarding privacy requirement 20 and the formula 21 enabling the satisfaction of the subscriber requirement on the Code Protection feature 16.

4.3 Feature representation in SDL

Feature Life Cycle To specify feature behaviours and their relationships with the Basic Call in SDL, the notion of *Feature Life Cycle* should be taken into account[21, 4]. This notion concerns the feature's behaviour when it is already deployed in the network and subscribed to by an end-user. Interactions between services can occur at different steps of the service life cycle so the following notions should be explicit in the SDL specification.

Parameterization where the end user modifies data related to the feature.

Activation where the subscriber activates the feature, i.e. prepares it for invocation.

Invocation where a request is received and the feature is executed.

Deactivation where the subscriber cancels the feature activation.

Data Modification where feature data is modified by execution of this, or another, feature.

Interleaving Execution Since the objective of our method is to detect the greatest number of logical feature interactions, we do not want to have to introduce implementation choices, such as an ordering of feature invocations, in our model. If the behaviour resulting from the invocation of several features depends on the order in which they are invoked, then we must be able to specify non-determinism in this ordering in order to detect such conflict interactions. Consider, for example, the OCS and ABD features where screening before or after translation can change the result. Many other examples could be cited (Call Waiting and Call Forwarding on Busy Line, Call Forwarding and Terminating Call Screening, ...). The lack of non-determinism of SDL88 means that each possible ordering has to be explicitly specified. For this reason, the use of SDL92 (or the slightly extended SDL88 used in the GEODE tool, see section 2.4) gives rise to more manageable specifications.

5 Conclusion

On existing Tools The model described in this paper has been developed for several IN service features. The application of the method to detect interactions was based on the GEODE/FV [3] verification tool, the model being specified in SDL and the properties with MSCs (Message Sequence Charts), and many detection of interactions by property violation have been obtained. Unfortunately, verification techniques using MSCs are not powerful enough for many of the properties of interest. However, the power of the tool will soon be increased to allow the expression of properties with SDL observers (automata). The derivation of observers from temporal logic formulas

can then enable verification of the temporal logic formulas on simulation runs. In the meantime, temporal logic formulas will be used as a documentation language.
Some service feature interactions cannot be detected by the property violation method outlined here. For example, an interaction which provokes a supplementary phenomenon not produced during the execution of each of the service features considered separately and therefore not foreseen, is such an interaction. In addition, some feature properties simply cannot be expressed using temporal logic, MSCs or SDL observers. In these cases only semi-automatic methods can be applied, using, in particular, animation tools but also with the help of some functionalities of state-space exploration tools: detection of deadlock or livelock and analysis of scenarios issued from the state-space exploration. Even for non-automatic techniques, formal specification is necessary to analyse correctly the real complex behaviour of network and services.

Service Feature Properties. A major challenge for a feature interaction detection method such as the one discussed in this paper, is to express *stable* service feature properties, i.e. properties which do not depend on the architecture model and the service features previously introduced. Another issue of importance for facilitating future integration of such methods in a service creation environment, is the necessity to introduce user-friendly macro-operators like the **SOON** operator introduced in this article.

References

[1] Computer, vol. 26, n. 8. IEEE Computer Society, August 1993.

[2] Ieee communications magazine, vol. 31, n. 8. IEEE Communications Society, August 1993.

[3] B. Algayres, Y. Lejeune, F. Hugonnet, and F. Hantz. The AVALON Project: A VALidatiON Environment for SDL/MSC Descriptions. In *Proc. 6th SDL Forum*, 1993.

[4] H. Blanchard, A.M. Daniel, J. Muller, J.M. Pageot, P. Razol, and L. Vallee. Le cycle de vie d'un service et son application aux etudes d'interaction. CNET: SERENITE/DT/INT/CYCL, 1993.

[5] W. Bouma and H. Zuidweg. Formal analysis of service/feature interaction using model checking. In *Feature Interaction Workshop, St. Petersburg*, December 1992.

[6] A. Bourguet-Rouger and P. Combes. Exhaustive validation and test generation in elvis. In *Proc. Fourth SDL forum*, North Holland, 1989.

[7] T.F. Bowen, F.S. Dworak, C.-H. Chow, N.D. Griffeth, G.E. Herman, and Y.-J. Lin. Views on the feature interaction problem. In *Proceedings of the 7th International Conference on Software Engineering for Telecommunications Switching Systems*, pages 59–62, Bournemouth, July 1989.

[8] A. Cavalli and L. Farinas Del Cerro. A deduction method for linear temporal logic. In *7th International Conference on Automated Deduction*, Napa, California, 1984.

[9] CCITT. New Recomendations Q1200 – Q series: Intelligent Network Recommendation. Technical report, CCITT, COM XI-R 210-E, 1992.

[10] E.M. Clarke, E.A. Emerson, M.C. Browne, and A.P. Sistla. Using temporal logic for automatic verification of finite state machine. Logics and Models of Concurrent Systems, 1985.

[11] P. Combes, M. Michel, and B. Renard. Formal verification of telecommunication service interactions using SDL methods and tools. In *Proceedings 6th SDL Forum*, Darmstadt, October 1993.

[12] P. Combes, B. Renard, W. Bouma, and H. Velthuijsen. Formalisation of properties for feature interaction detection. In *International Conference on Intelligence in Services and Networks*, Paris, France, November 1993.

[13] E.A. Emerson and J.Y. Halpern. Sometimes and not never revisited: On branching versus linear time temporal logic, 1986.

[14] E.A. Emerson and J. Srinivasan. Branching time temporal logic. In J.W. de Bakker, W.P. de Roever, and G. Rozenberg, editors, *Linear Time, Branching Time and Partial Order in Logics and Models for Concurrency*, number 354 in Lecture Notes in Computer Science, pages 123–172. Springer Verlag, Berlin, 1989.

[15] V. Encontre. Geode : An industrial environment for designing real time distributed systems in sdl. In *Proc. Fourth SDL forum*, North Holland, 1989.

[16] S. Graf, J.L. Richier, C. Rodriguez, and J. Voiron. What are the limits of model checking methods for the verification of real life protocols?. In Springer Verlag, editor, *Workshop on Automatic Verification Methods for Finite State Machines*, 1989.

[17] N.D. Griffeth and H. Velthuijsen. The negotiating agent model for rapid feature development. In *Proceedings of the Eighth International Conference on Software Engineering for Telecommunications Systems and Services*, Florence, Italy, March/April 1992.

[18] F. Kroger. *Temporal Logic of Programs*. EATCS, Monographs on Theoretical Computer Science. Springer Verlag, 1987.

[19] Z. Manna and A. Pnueli. *The Temporal Logic of Reactive and Concurrent Systems*. Springer Verlag, 1992.

[20] J. Muller, H. Blanchard, P. Combes, A.M. Daniel, and J.M. Pageot. Perfection is not of this world: debating a user-driven approach of interaction in an advanced intelligent network. In *Proc. TINA 1993*, 1993.

[21] J. Muller, P. Razol, and E. Bac. Some requirements and techniques for service interaction in an intelligent network. In *Proc. TINA 1991*, 1991.

[22] J.L. Richier, C. Rodriguez, J. Sifakis, and J. Voiron. Verification in xesar of the sliding window protocol. In *Proc. 7th International Workshop on Protocol Specification and Validation*, North Holland, 1987.

[23] J. Rumbaugh, M. Blaha, W. Premerlain, F. Eddy, and W. Lorenson. *Object-Oriented Modeling and Design*. Prentice Hall, 1991.

[24] SPECS-SEMANTICS and ANALYSIS. Final definition of a property language on the common semantics. Deliverable 46/SPE/WP5/DS/013/b1, SPECS, 1992.

[25] C. Stirling. Modal and temporal logics. ECS-LFCS-91-157, 1991.

Specifying Features and Analysing Their Interactions in a LOTOS Environment

M. Faci and L. Logrippo

*University of Ottawa, Protocols Research Group,
Department of Computer Science, Ottawa, Ontario, Canada K1N 6N5
E-mail: {mfaci, luigi}@csi.uottawa.ca*

Abstract. This paper presents an approach for specifying telephone features and analysing their interactions in a LOTOS environment. The approach is characterized by a flexible specification structure and an analysis method based on knowledge goals. Structurally, the specifications allow the integration of new features into existing ones by specifying each feature independently and composing its behaviour with the existing system. Analytically, the reasoning mechanism allows the specifier to analyse features, for the purpose of detecting their interactions, by defining knowledge goals and simulating the system to verify if they are reachable. A non reachable knowledge goal reveals the existence of a feature interaction, or a design error. We explain this approach by the use of two, now classical, examples of feature interactions, namely Call Waiting & Three Way Calling and Call Waiting & Call Forward on Busy.

Keywords: Telephone Features, Feature Interactions, Formal Specifications, LOTOS.

1. Motivation and Background

The plain old telephone service (POTS) is used for establishing a communication session between two users. A telephone feature, such as *call waiting* (cw), *call forward on busy* (cfb), and *three way calling* (3wc), is defined as an added functionality of POTS. Augmenting POTS with a small set of features is considered to be a technically straightforward job. Both the behaviours of POTS and the features are analysed and decisions are taken as to how to integrate the features into POTS. Conflicts between any of the features are resolved on a case by case basis. However, as more and more features need to be integrated, as is the case for present and future telephone networks [Lata89], the task becomes more difficult. Features that perform their functions satisfactorily on their own are, in some instances, prevented from doing so in the presence of other features. This problem has become known as the *feature interaction problem* [BDCG89].

Investigations into the feature interaction problem fall into one of three complementary categories[CaVe93]: Detection, avoidance, and resolution. The objective of a detection approach is to analyse a set of independently specified features and determine whether or not there are any conflicts between their joint behaviour [CaLi91], [Lee92], [BoLo93], [DaNa93]. An avoidance mechanism assumes that the causes of the interactions are known and an architectural or analytical approach is defined to prevent the manifestation of such interactions [MiTJ93]. The avoidance approach is most suitable in the early phases of specification and design of features. Finally, the objective of a resolution mechanism is to find appropriate solutions to interactions that manifest themselves at execution time [Cain92] [GrVe92].

This paper describes a method, based on the LOTOS specification language [BoBr87], [LoFH92], for detecting feature interactions at the specification level, in the context of single user single element features[CGLN93]. Central to our method are the concepts of *constraints* and *knowledge goals*. Constraints, which we have categorized as local, end-to-end, and global, are used to structure the specification so that new features can be added to the specification or removed from it with plausible ease [FaLS91]. While the concept of constraints is useful for structuring specifications, the concept of knowledge goals is useful for reasoning about telephone features, in order to detect feature interactions. It is based on the theories of *knowledge* which are being developed for understanding and reasoning about communication protocols and distributed systems [HaFa89], [HaMo90], [PaTa92]. In the knowledge-based approach, the evolution of the system can be described by the evolving knowledge of the components, about the state of other components, which collectively make up the global state. Thus, the state of knowledge in the system changes as a result of exchanging messages between the communicating components, and communication between the components depends on their knowledge about the system state. We use this concept to analyse combinations of features in order to detect the interactions between them.

In section 2, we describe our method for detecting interactions between independently specified features. In section 3, we demonstrate the application of our approach on two examples: *cw&cfb* and *cw&3wc*. We conclude with some thoughts regarding our research directions in section 4.

2. Using a LOTOS Environment to Detect Feature Interactions

Although the feature interaction problem has existed for quite many years, little attention was paid to it until it was explicitly defined [BDCG89]. Since then, a whole new field of interest is born. Due to the lack of space, we simply point the reader to some of the work of other researchers in this area. In particular, we mention the work of [HoSi88], [CaLi91], [Dwor91], [Cain92], [EKDB92], [GrVe92], [Inoue92], [Lee92], [DaNa93], and [Zave93]. Two other excellent sources of information are the special issues of IEEE Computer[Comp93] and IEEE Communications magazine [Magz93].

We begin the section by reviewing the constraint-oriented style that we had developed for specifying telephone systems in LOTOS. Then, we describe an improvement to the structure of our specifications which makes it possible to easily integrate features into a telephone system, while still using the concept of constraints. We conclude the section by describing a reasoning mechanism which allows us to analyse the joint behaviour of features, for the purpose of detecting their interactions.

2.1 LOTOS Structure of the POTS Specification: An Overview

We have previously developed a LOTOS specification structure that is well suited for specifying the behaviour of telephone systems[FaLS91]. The structure is based on the constraint-oriented specification style [VSVB91], where we identified three types of constraints:

(1) *Local constraints* are used to enforce the appropriate sequences of events at each telephone, and are different according to whether the telephone is a *Caller* or a *Called*. Therefore local constraints are represented by the processes *Caller* and *Called* and an instance of each of these is associated with each telephone existing in the system. Because these two processes are independent of each other, they are composed by the interleaving operator |||.

(2) *End-to-End* constraints are related to each connection, and enforce the appropriate sequence of actions between telephones in a connection. For example, ringing at the *Called* must necessarily follow dialling at the *Caller*. Process *Controller* enforces these constraints. Because they must apply to both *Caller* and *Called,* we have the structure *(Caller ||| Called) || Controller*. Thus the controller must participate in every action of the *Caller*, as well as in every action of the *Called*, separately.

(3) *Global constraints* are system-wide constraints. In the POTS context, we identified one main constraint, which is the fact that at any time, a number is associated with at most one connection. Because global constraints, represented by a process *GlobalConstraints,* must be satisfied simultaneously over the whole system, we have the structure *Connections || GlobalConstraints*.

It should be stressed that the constraint-oriented style is purely a specification style, which allows to clearly separate the logical constraints a system must abide. It does not necessarily reflect an implementation architecture. To obtain an implementation architecture from a constraint-oriented structure, style transformations may be applied [VSVB91].

2.2 Enhancing the Structure of POTS Specifications

A graphical representation of the enhanced structure of POTS specifications is shown in Fig.1. The LOTOS specification of this structure, which handles only 4 users for the purposes of illustration, is shown in Fig. 2. This structure has the following characteristics: (1) Each user is represented by two processes, a caller side and a called side and each caller process is bound to its own controller. So, in the POTS case, each connection becomes: *(Caller(n)* ||| *Called)* || *Controller(n)*, where *n* is the subscriber number. (2) Contrary to the structure presented in [FaLS91], where the number of the called user was passed to the called process at instantiation time, in this structure, the called process is now represented as a set of alternatives, where each alternative represents the called side of a user in the system. Clearly, since the controller offers to synchronize with only one called at a time, only a single alternative will offer the same value. As we will see in section 3.1, this structure is highly flexible for integrating new features into a telephone system. (3) the global constraints process participates in every action in which any instance of the controllers participates.

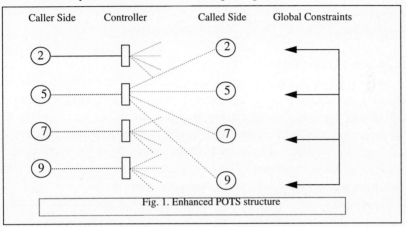

Fig. 1. Enhanced POTS structure

```
1   behaviour
2   (   ( Caller[Suser] (2) ||| Called [Suser](Users) )|| Controller [Suser] (2)
3       |||
4       ( Caller[Suser] (5) ||| Called [Suser](Users) ) || Controller [Suser] (5)
        |||
5       (Caller[Suser] (7) ||| Called [Suser](Users) ) || Controller [Suser] (7)
6       |||
7       ( Caller[Suser] (9) ||| Called [Suser](Users) ) || Controller [Suser] (9)   )
8   ||
9   GlobalConstraints[Suser](parameters)
10  where
11      process Caller[Suser](n: TelNo):noexit:= ...
12      process Called [Suser](Users: List): noexit:=
13          Called [Suser](2) [] Called [Suser](5) [] Called [Suser](7) [] Called [Suser] (9)
14      endproc (* Called *)
15      process Controller[Suser](n: Digit): noexit:= ...
16      process GlobalConstraints [Suser](Parameters: Sets)): noexit:= ...
```

Fig. 2. LOTOS specification of POTS using the enhanced structure.

2.2.1 Local Constraints

As mentioned, *Local constraints* are used to enforce the appropriate sequences of events within each process. For example, to specify a *Caller* process, the specifier needs only to understand the events that are exchanged between the *Caller* process and its environment. At this stage, the specifier does not need to concentrate on *who* represents the environment or *which* processes will interact with the *Caller* process. These concerns are addressed at a later stage of the specification. By taking this view, we have in fact reinforced the concept of *separation of concerns*. In the case of specifications dealing with POTS [FaLS91] [BoLo93], *local constraints* were applied to the caller entity and the called entity only. Our experience has shown that the concept of local constraints can be used for specifying telephone features as well.

2.2.2 End-to-End Constraints

For a simple two-way call processing, the *end-to-end* constraints were used to synchronize the actions of two processes with respect to each other, most often the sender and the receiver. Our experience in writing the POTS specifications is that establishing a temporal order between two actions, one being offered by the caller and the other one by the called, is quite intuitive and simple. However, we recognize that expressing *end-to-end* constraints of the new structure may become more complicated, because there are more processes which may offer synchronization actions. And, it is still the specifier's task to impose a temporal order on a set of given actions. The specifier must then have some heuristics and guidelines at his/her disposal. A possible approach is to start by expressing the end-to-end constraints as *cause-effect*[Lin90] [NuPr93] rules.

2.2.3 Global Constraints

Finally, the *Global constraints* are at a higher level of abstraction than the end-to-end constraints, since they are imposed on the global behaviour of the system. In the simple two-way call processing model, the global constraints were restricted to enforcing *value* constraints between independent connections. In our new structure, global constraints gain an added importance. They enforce *control* constraints as well. Let us illustrate this with the following example, see Fig. 4. Suppose that user 2, who subscribes to *cw*, establishes a connection (A) with user 5. Also, suppose that, while 2 is talking to 5, 7 calls 2. Since 2 has *cw*, the global constraints process must manage the new connection (B), which consists of a new caller and shares the called side with the existing connection of 2. Therefore, the global constraint is responsible for switching *the control* between connection A and connection B, depending on what stage of the communication the users are in. For instance, when 7 dials 2 and 2 answers the call, the global constraint removes the communications between 5 and 2 from the set of active sessions and inserts it into the set of holding sessions. At the same time, it allows a communication session to be established between 7 and 2. When 2 and 7 finish their conversation, the global constraint reactivates the connection (A), between 2 and 5.

The structure that we have just defined exhibits the required flexibility. New features are defined in terms of their local constraints and their end-to-end constraints, while their global constraints are composed with the existing global constraints of the system. To specify a new feature, the specifier either instantiates actions that already exist in the system, such as the *dial* action, or defines new actions, such as the *CwTone* action used in the definition of the *cw* feature. In the former case, the resulting global constraints on the *dial* action is the conjunction of the existing constraints and the new constraints. In the latter case, a new alternative action is added to the behaviour of the global constraint process.

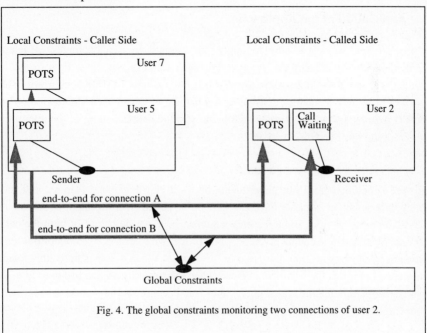

Fig. 4. The global constraints monitoring two connections of user 2.

2.3 Using Knowledge Goals to Reason about LOTOS Specifications

This section describes how to adapt the *knowledge-oriented* model of Halpern and Moses [HaMo90] and incorporate it into LOTOS specifications for telephony systems. The intuition we want to capture, by using the knowledge-based approach, is that the designer reasons about LOTOS processes in terms of how relevant information, from the local point of view (i.e., local constraints) of each process, becomes satisfied at certain points during the execution of the system. To analyze whether two features, say *cw* and *cfb*, interfere with each other, the designer defines a set of knowledge goals and verify their reachability, when both features are active. If any of the goals is unreachable, the designer concludes that a feature interaction (or design error) exists. Otherwise, no conclusion can be drawn from the analysis. In a way, this is similar to system testing. A test which fails to reveal an error does not indicate that the system under test is error free, it only means that the system is error free with respect to the assumption expressed by the test. Details of our analysis using two examples are given in section 3.

It is interesting to emphasize that each process reasons about the outside world only in terms of its local information. Therefore, a process moves from one state to another state based only on its knowledge. Similarly, it gains (or loses) new knowledge as it moves from one state to another. Also, notice that knowledge in this context is an *external* notion, in the sense that processes do not acquire knowledge on their own nor are they able to analyze the knowledge state of other processes. For the purposes of analysis, the specifier is responsible for choosing the appropriate *knowledge goals* used to reason about the system.

3. Application of the Approach

In this section, we show how to apply our method to two examples of feature interactions: *cw* vs. *cfb* and *cw* vs. *3wc*. For each example, we show how to use the enhanced structure to integrate the two features into a single LOTOS specification, and then we show how to use the reasoning mechanism to detect their interactions. Only LOTOS segments which contribute significantly to the understanding of the integration and reasoning mechanisms are given.

3.1 Specifying Features in LOTOS

We have concluded from our experiments of specifying telephone features that most features act on behalf of either the caller side or the called side. From our structural and analytical points of view, some features which seem to act on behalf of both the caller side and the called side can be given only one of the two roles. An example of this is the *automatic recall* feature (not discussed in this paper). When a user is busy, this feature automatically returns the last incoming call when the subscriber's line becomes idle. We have classified this feature as having a caller role because its first action is to initiate a connection back to a user who has initiated a call.

3.1.1 Call Waiting

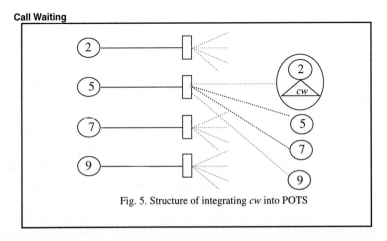

Fig. 5. Structure of integrating *cw* into POTS

Call Waiting is a feature which generates a call waiting tone, to alert a busy user that a second incoming call is waiting to be answered. The user may choose to answer the call, using a special *flashhook* signal, or may simply continue with the original communication and ignore the call waiting tone.

This feature has a called role. Therefore, first, we specify the behaviour of the feature in the context of POTS. Second, we modify the existing *Controller* so that it handles any new actions that the new feature participates in, such as *call waiting tone*. This results in a structure of the form: *(Caller(n) ||| Called) || ControllerCw(n)*. Third, we replace the existing definition of the controller with the new definition, which is now capable of handling POTS calls as well as subscribers with the call waiting feature. Finally, for each action in which the feature participates, we check that the predicates remain valid in the global constraints. Modifications of figure 1 are shown in Fig. 5 and result in the structure of Fig. 6.

```
1    behaviour
2    (   ( Caller[Suser] (2) ||| Called [Suser](Users) ) || Controller [Suser] (2)
3        |||
4        ( Caller[Suser] (5) ||| Called [Suser](Users) ) || Controller [Suser] (5)

5        |||
6        ( Caller[Suser] (7) ||| Called [Suser](Users) ) || Controller [Suser] (7)
7        |||
8        ( Caller[Suser] (9) ||| Called [Suser](Users) ) || Controller [Suser] (9)    )
9    ||
10   GlobalConstraints[Suser](parameters)
11   where
12       process Caller[Suser](n: TelNo):noexit:= ...
13       process Called [Suser](Users: List): noexit:=
14           CalledPots [Suser](2) [] CalledPots [Suser](5) [] CalledPots [Suser](7)
15           [] CalledPots [Suser] (9)
16           [] CalledCw(2)                                   <---------------Added
17       endproc (* Called *)
18       process Controller[Suser](n: Digit): noexit:= ... <---------------- Modified
19       process GlobalConstraints [Suser](Parameters: Sets)): noexit:= ... <--- Modified
```

Fig. 6. LOTOS specification of *cw* within POTS.

3.1.2 Call Forward on Busy
Call Forward on Busy is a feature which allows a user, who is already involved in a conversation with a second user, to transfer his/her incoming calls to a predetermined third user. Depending on the specifier's intentions, the busy user may or may not be informed that a call transfer has occurred. This feature acts on behalf of the called side, so its specification is similar to that of call waiting. The modification of Fig. 1 results in a similar structure to that of call waiting, and is not shown.

3.1.3 Three Way Calling
Three way calling is a feature which allows a user, who is already involved in a conversation with a second user, to add a third user to the conversation. The subscriber

of the feature must put the second user on hold, using a special *flashhook* signal, while establishing a communication with the third user. Once the communication is established, a second *flashhook* brings the second user back to the conversation to form a 3 way communication.

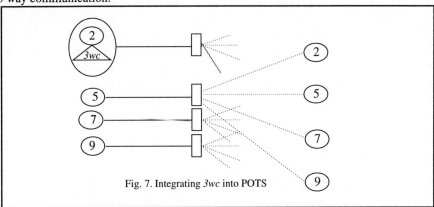

Fig. 7. Integrating *3wc* into POTS

This feature has a caller role. It is specified as follows. First, we specify the behaviour of the feature with respect to POTS. Second, we modify the existing *Controller* so that it handles any new actions that *3wc* participates in, such as *flashhook*. This results in a structure of the form: *(Twc(n)* ||| *Called)* || *Controller3wc(n)*, where *n* identifies the subscriber for which the feature is to be invoked. Third, we compose this new structure with the existing connections using the ||| operator. Finally, for each action in which the feature participates, we check that the predicates in the global constraints process remain valid. Extending our specification of Fig. 1, the structures of Figs. 7 and 8 result:

```
1   behaviour
2   (   ( Caller[Suser] (2) ||| Called [Suser](Users) ) || Controller [Suser] (2)
3       |||
4       ( Caller3wc[Suser] (2) ||| Called [Suser](Users) ) || Controller3wc [Suser](2) <-------- added
5       |||
6       ( Caller[Suser] (5) ||| Called [Suser](Users) ) || Controller [Suser] (5)
7       |||
        ( Caller[Suser] (7) ||| Called [Suser](Users) ) || Controller [Suser] (7)
8       |||
9       ( Caller[Suser] (9) ||| Called [Suser](Users) ) || Controller [Suser] (9)
10  )
11  ||
12  GlobalConstraints[Suser](parameters)
13  where
14          process Caller[Suser](n: TelNo):noexit:= ...
15          process Caller[Suser](n: TelNo):noexit:= ...
16          process Called [Suser](Users: List): noexit:=
17              Called [Suser](2) [] Called [Suser](5) [] Called [Suser](7) [] Called [Suser] (9)
18          endproc (* Called *)
19          process Controller[Suser](n: Digit): noexit:= ...
20          process Controller3wc [Suser](n: Digit): noexit:= ... <-------- Added
21          process GlobalConstraints [Suser](Parameters: Sets)): noexit:= ...
```

Fig. 8. New LOTOS Structure of POTS

3.2 Analysing Features to Detect their Interactions

In this section, we present our method and show how it can be used to detect interactions between *cw&cfb* and *cw&3wc*. The method calls for the following steps to be carried out:

1• Specify each feature independently, within a POTS context;

2• Use the structure defined in section 2.2 to integrate both features into a single specification;

3• Define the knowledge goals to be reached in the reasoning phase; a knowledge goal is expressed as a LOTOS process which is composed in parallel with the specification obtained in 2 above.

4• Finally, simulate the system and check if the selected goals are reachable. A feature interaction (or design error) is detected if the selected goals are not reachable.

Assuming that points 1 and 2 above are completed successfully, let us proceed with point 3. In the two examples to follow, the designer has chosen to reason about the system in terms of subscribers' talking states. In other words, the designer knows that each feature, when active by itself within the context of POTS, can successfully reach its talking state, and the question she is attempting to answer is: can two features, if activated simultaneously by the same user, reach their talking states? as will be seen, the answer is no in both examples. Notice that if the designer fails to detect an interaction using a selected knowledge goal, it is strongly recommended that other goals be tried. So, it is the responsibility of the designer, using her design insights, to define the appropriate goals for reasoning about the system.

3.2.1 Call Waiting & Call Forward on Busy

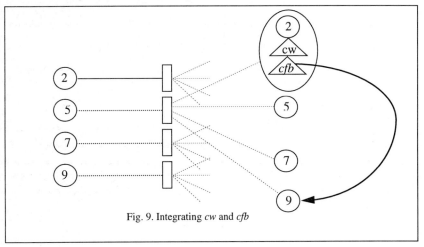

Fig. 9. Integrating *cw* and *cfb*

Fig. 10 shows the modification of POTS by extending its functionality according to *cw* and *cfb*. Let us assume that only cw is active on 2. Then, if 7 calls 2 while 2 is talking to 5, one possible scenario is that 5 is put on hold and a talking session between 7 and 2 is established. This scenario is shown on the left side of Fig. 10. If we assume instead that only cfb is active, and using the same execution scenario, a talking session between 2 and 5 remains active and a talking session is established between 7 and 9. Therefore, our knowledge goal is defined as: Talk(2, 7) *and* Talk(7, 9). Since each of these two goals is reachable when only one feature is active, then they *must* also be reachable if cw and cfb are activated simultaneously with respect to their talking sessions.

To clarify this further, we denote by POTS+CW+CFB[gpots, gcw, gcfb] the extension of POTS by the features *cw* and *cfb*, where each feature is added to POTS separately; *gpots* is the gate through which all POTS events occur, *gcw* is the gate through which all *cw* events occur, *gcfb* is the gate through which all *cfb* events occur. Let the first part of the goal Talk[gpots, gcw](2, 7) be defined by the expression (gpots !7 !dials !2; gcw !2 !flashhook), as shown in the left branch of Fig. 10; let the second part of the goal Talk[gpots, gcfb](7, 9) be defined by the expression (gpots !7 !dials !2; gcfb

!9 !rings (from 7); gcfb !9 !answers), as shown in the right branch of Fig. 10. Therefore, if a deadlock (in the sense of LOTOS) occurs in the behaviour expression:
POTS+CW+CFB[gpots, gcw, gcfb] || (Talk[gpots, gcw] |[gpots]| Talk[gpots, gcfb]),
then we can conclude that a feature interaction exists. The LOTOS specification and its execution tree are given as follows:

```
3    behaviour
4         gpots !7 !dials !2;
5         (  gcw !2 !flashhook; stop
6         []
7            gcfb !9 !rings; gcfb !9 !answers !7; stop
8         )
9         ||
10        (
11           gpots !7 !dials !2; gcw !2 !flashhook; stop
12           |[gpots]|
13           gpots !7 !dials !2; gcfb !9 !rings; gcfb !9 !answers !7; stop
14        )
```

Execution tree
    1 gpots [4,11,13]
        | 1 gcw [5,11] **DEADLOCK**
        | 2 gcfb [7,13] **DEADLOCK**

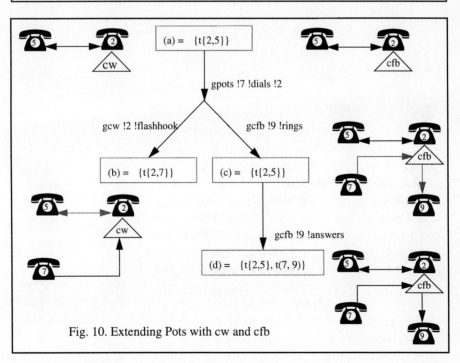

Fig. 10. Extending Pots with cw and cfb

3.2.2 Call Waiting & Three Way Calling

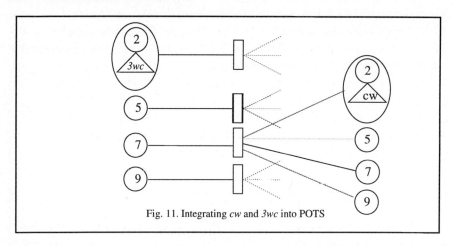

Fig. 11. Integrating *cw* and *3wc* into POTS

Our analysis for detecting the interaction between *cw* and *3wc* is similar to that of the previous example. Clearly, our objective is to know whether or not *cw* and *3wc* can be invoked simultaneously by 2 without causing a conflict between them.

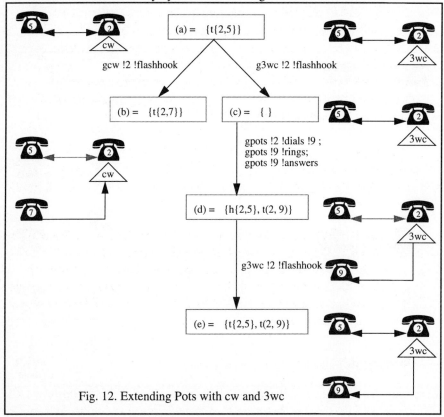

Fig. 12. Extending Pots with cw and 3wc

The left side of Fig. 12 is the same as the left side of Fig. 10. Concerning the right-hand side, assume that only 3wc is active, and that, using the same flashhook signal, 2 moves to a state where he may dial 9. To make the analysis easier, we assume that 9 is idle and answers the call. Therefore, a talking session between 2 and 9 is established. A second flashhook reestablishes the original talking session between 2 and 5. Therefore, our knowledge goal can be defined as: Talk(2, 7) *and* Talk(2, 9). As is in the previous example, we must show that the behaviour: POTS+CW+3WC[gpots, gcw, g3wc] || (Talk[gpots, gcw] |[gpots]| Talk[gpots, g3cw])) is deadlock free. In this case as well, a deadlock is reached as can be easily verified by executing the above behaviour expression. We conclude that a feature interaction exists.

4. Conclusions and Research Directions

We have proposed a method, based on a formal approach, for detecting feature interactions at the specification level, for single user single element features. Structurally, the approach uses three types of constraints: local constraints, end-to-end constraints, and global constraints. This structure allows the integration of new feature specifications into existing ones simply by classifying their roles as *caller* or *called* and expressing their global constraints as a conjunction. This structure is possible because of LOTOS's multiway synchronization mechanism, which also offers the flexibility to describe a system as a composition of constraints. Analytically, the approach is based on a reasoning mechanism which allows the specifier to analyse features, for the purpose of detecting their interactions, based on knowledge goals, where each goal is expressed as a LOTOS process. Our interpretation, in this context, is that conflicts between features correspond to deadlock situations in the LOTOS sense.

Important items for future research are adapting our approach to more realistic examples, extending the technique to other types of feature interactions, and making the technique more automatic. i.e., less dependent on designer's insight regarding where the problem might be found.

Acknowledgment. Funding sources for our work include the Natural Sciences and Engineering Research Council of Canada, the Telecommunications Research Institute of Ontario, Bellcore, Bell-Northern Research, and the Canadian Department of Communications. We like to acknowledge the many fruitful discussions that we had with Bernard Stepien and members of our LOTOS group. Also, comments from the referees have led to improvements of the content of this paper.

5. References

[BDCG89]   T.F. Bowen, F.S. Dworak, C.H. Chow, N. Griffeth, G.E. Herman, and Y-J. Lin, The Feature Interaction Problem in Telecommunications Systems, 7th International Conference on Software Engineering for Telecommunication

Switching Systems, July 1989, 59-62.

[BoBr87] Bolognesi, B.,Brinksma, E. Introduction to the ISO Specification Language LOTOS. Computer Networks and ISDN Systems 14, 1987, 25-59.

[BoLo93] R. Boumezbeur, L. Logrippo, Specifying Telephone Systems in LOTOS, IEEE Communications Magazine, Aug. 1993, 38-45. E. J. Cameron, N. Griffeth, Y. Lin, M. E. Nilson, W. K. Schnure, H. Velthuijsen, A Feature Interaction Benchmark for IN and Beyond, IEEE Communications, vol. 31, No. 3, 64-69, March 1993.

[CaLi91] E. J. Cameron and Y.J. Lin, A Real-Time Transition Model for Analyzing Behavioral Compatibility of Telecommunications Services. In Proceedings of the ACM SIGSOFT 1991 Conference on Software for Critical Systems, pp. 101-111, December 1991, New Orleans, Louisiana.

[CaVe93] J. Cameron and H. Velthuijsen, Feature Interactions in Telecommunications Systems, IEEE Communications Magazine, Aug. 1993, 18-23.

[Cain92] M. Cain, Managing Run-Time Interactions Between Call-Processing Features, IEEE Communications Magazine, pp. 44-50, February 1992.

[Comp93] IEEE Computer, Special Issue on Feature Interactions in Telecomunications Systems, Aug. 1993.

[DaNa93] O. Dahl and E. Najm, Specification and Detection of IN Service Interference Using LOTOS, to appear in the proceedings of Forte '93, Boston.

[Dwor91] F. S. Dworak, Approaches to Detecting and Resolving Feature Interactions, GLOBECOM 1991, pp. 1371-1377.

[EKDB92] M. Erradi, F. Khendek, R. Dsouli, and G. V. Bochmann, Dynamic Extension of Object-Oriented Distributed System Specifications, First International Workshop on Feature Interactions in Telecommunications Software Systems, Florida, 1992, 116-132.

[FaLS91] Faci, L. Logrippo and B. Stepien,Formal Specifications of Telephone Systems in LOTOS: The Constraint-Oriented Style Approach, Computer Networks and ISDN Systems, 21, 52-67, North Holland, 1991.

[GrVe92] N. D. Griffeth and H. Velthuijsen, The negotiating agent model for Rapid Feature Development, Proceedings of the 8th International Conference on Software Engineering for Telecommunications Systems and Services, Florence, Italy, March/April 1992.

[HaFa89] J. Y. Halpern and R. Fagin, Modelling Knowledge and action in distributed systems, Distributed Computing, 3, 159-177, 1989.

[HaMo90] J. Y. Halpern and Y. Moses, Knowledge and Comon Knowledge in a Distributed Environment, JACM, Vol. 37, No. 3, 549-587, July 1990.

[HoSi88] S. Homayoon and H. Singh, Methods of Addressing the Interactions of Intelligent Network Services With Embedded Switch Services, IEEE Communications Magazine, pp. 42-47, Dec. 1988.

[Inoue92] Y. Inoue, K. Takami, and T. Ohta, Method for Supporting Detection and Elimination of Feature Interaction in a Telecommunication System, First International Workshop on Feature Interactions in Telecommunications

Software Systems, Florida, 1992, 61-81.

[Lata89] LATA Switching Systems Generic Requirements (LSSGR), Bellcore, TR-TSY-000064, FSD 00-00-0100, July 1989.

[Lee92] A. Lee, Formal Specification and Analysis of Intelligent Network Services and their Interaction, Ph. D. Thesis, Dept. of Computer Science, University of Queensland, 1993.

[Lin90] Y. J. Lin, Analyzing Service Specifications Based upon the Logic Programming Paradigm, In Proceedings of the IEEE GLOBECOM 1990, pp. 651-655, December 1990, San Diego, California.

[LoFH92] L. Logrippo, M. Faci and M. Haj-Hussein, An Introduction to LOTOS: Learning by Examples, Computer Networks & ISDN Systems, Vol. 23, No. 5, 1992, pp. 325-342.

[Magz93] IEEE Communications Magazine, Special Issue on Feature Interactions in Telecomunications Systems, Aug. 1993.

[MiTJ93] J. Mierop, S. Tax, R. Janmaat, Service Interaction in an Object Oriented Environment, IEEE Communications Magazine, Aug. 1993.

[NuPr93] K. Nursimulu and R. L. Probert, Cause-Effect Validation of Telecommunications Service Requirements, Technical Report TR-93-15, University of Ottawa, Dept. of Computer Science, October 1993.

[PaTa92] P. Panangaden and K. Taylor, Concurrent Common Knowledge: Defining Agreement for Asynchronous Systems, Distributed Computing, 6, 73-93, 1992.

[VSVB91] C.A. Vissers, G. Scollo, M. van Sinderen, E. Brinksma, Specification Styles in Distributed Systems Design and Verification, Theoretical Computer Science 89, 1991, 179-206.

[Zave93] P. Zave, Feature Interactions and Formal Specifications in Telecommunications, IEEE Computer Magazine, Aug. 1993, 18-23.

Towards a Formal Model for Incremental Service Specification and Interaction Management Support

Kong Eng CHENG
Centre for Advanced Technology in Telecommunications,
Collaborative Information Technology Research Institute
and
Department of Computer Science, Royal Melbourne Institute of Technology
GPO Box 2476V, Melbourne, Victoria 3001, Australia
Email: kec@cs.rmit.edu.au

Abstract. This paper presents a technique for incremental service specification and interaction management. It introduces a mechanism for linking services, known as behaviour chaining to support independent service specification and service composition. The constraint oriented style of specification is used to manage interaction. The overall architecture is formally specified using LOTOS.

1. Introduction

This paper presents a technique for incremental service specification and interaction management. It introduces the concept of behaviour chaining to support incremental specification and addition of services. The constraint oriented style of specification [1] is used to manage service interaction. The approach is formally specified using LOTOS [1] and illustrated using a simplified process graph presented in [2].

The main idea is to allow each new service to be specified independently based on a common call model and/or existing services/service features (S/SFs). This is achieved by incrementally chaining existing or new S/SF behaviours together to develop the required new service. Process composition [1] is used as the underlying mechanism to represent the process of chaining S/SF behaviours to develop new services. Constraints may be applied during process composition to manage and resolve service interaction.

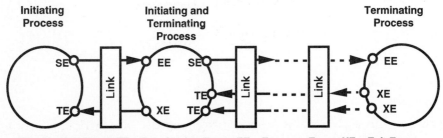

SE = Starting Event   TE = Terminating Event   EE = Entrance Event   XE = Exit Event

Figure 1. Behaviour chaining. It is assumed that a path exists between the entrance event and the exit events of each terminating process.

2. Incremental Services/Service Features Specification

2.1. Specification Approach - Behaviour Chaining

As shown in Figure 1, the overall specification approach consists of a sequence of process behaviours interconnected by links. The process behaviour models an object of interest such as a call model or a service behaviour. The link establishes a control and/or data relationship (interaction) between two or more process behaviours. Constraints may be applied to these links to restrict interaction between process behaviours. Each process behaviour can be further decomposed in a similar manner (refer to section 4).

An initiating process is defined as the process that starts the chaining activity while a terminating process is the process to be chained. A process can be an initiating and/or a terminating process. Each terminating process must define an entrance event (EE) where it can be linked into a chain and one or more exit events (XE) where the terminating process terminates at the chain.

The chaining process is performed by defining a starting event (SE) in the initiating process where the chain begins and one or more terminating events (TE) where the chain is suppose to end. The starting event and terminating events in the initiating process are linked to the entrance event and exit events of the terminating process, respectively through a link process as shown in Figure 1. Therefore, the links between process behaviours define the **interactions** that occur within the specification and we can concentrate on these links for analysis of interaction.

2.2. A simplified Example

As an illustration we use part of the simplified Basic Call State Model (BCSM) recommended in Q.1214 [3] as the basic call model. The originating BCSM (and similarly the terminating BCSM) is modelled as a LOTOS process and is represented as a process graph shown in Figure 2. The Points in Call (PICs) and Detection Points (DPs) [3] are represented as events along the arc. The dotted lines represent additional PICs and DPs that have not been specified.

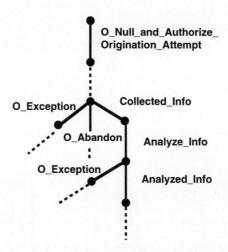

Figure 2. Originating BCSM. The direction of flow is top down unless indicated otherwise.

Events shown in BCSM in Figure 2 can be used to synchronise with other processes to introduce new services. A new service can be added (through behaviour chaining), without interfering directly with BCSM and existing services. This technique allows independent specification of individual services, and yet it provides a composition mechanism (through combinators in process algebra model) to integrate services together.

For instance to introduce call screening at the beginning of a call, the invocation of the service can be activated at an appropriate BCSM event and the control is returned to a suitable BCSM event upon completion. In this case, the invocation can be synchronised at the Analyze_Info event and the point of return is Analyzed_Info.

Assuming that the screening service is already defined, the terminating process (Screening_Service in Figure 3) must specify the entrance event (in this case Screening_Entrance) and the exit events (in this case Screening_Fail and Screening_Success) so that it can be associated by the link specify in Figure 4.

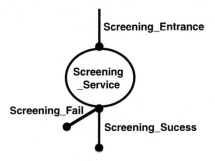

Figure 3. Defining entrance and exit events for a service. In this case the entrance event is Screening_Entrance and the exit events are Screening_Fail and Screening_Success.

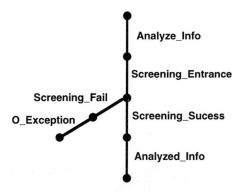

Figure 4. Link - chaining screening service to BCSM.

By parallel composition (in LOTOS terminology) of the processes in Figures 2, 3 and 4, based on the Analyze_Info and Analyzed_Info events, the combined behaviour is shown in Figure 5.

The composition process can be interpreted as attaching more 'branches' (ie. add new behaviours) to the main BCSM graph. This approach to compositionality is useful in supporting a systematic addition of new services without actually modifying ('touching') the BCSM. It allows new services to be specified independently and yet they can be easily integrated into existing services.

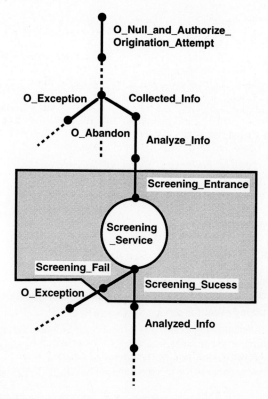

Figure 5. Composition of BCSM and service behaviour.

The following partial LOTOS specification formalises the above example:

```
/*
Initiating process     = BCSM
Starting event         = Analyze_Info
Terminating events     = O_Exception, Analyzed_Info
Terminating process    = Screening_Service
Entrance event         = Screening_Entrance
Exit events            = Screening_Fail, Screening_Success
Link =    Analyze_Info   -> Screening_Entrance
          O_Exception    -> Screening_Fail
          Analyzed_Info  -> Screening_Success
*/
BCSM [Analyze_Info, Analyzed_Info, O_Exception, ...]
| [Analyze_Info, Analyzed_Info, O_Exception] |
( Screening_Service_Link [Analyze_Info, Analyzed_Info, O_Exception,
   Screening_Entrance, Screening_Fail, Screening_Success]
  | [Screening_Entrance, Screening_Fail, Screening_Success ] |
  Screening_Service [Screening_Entrance, Screening_Fail,
   Screening_Success ]

  where
      process Screening_Service_Link [Analyze_Info, Analyzed_Info,
           O_Exception, Screening_Entrance, Screening_Fail, Screening_Success]
          Analyze_Info; Screening_Entrance;
```

```
    (  Screening_Fail; O_Exception; exit
    []  Screening_Success; Analyzed_Info; exit
    )
endprocess

process Screening_Service [Screening_Entrance, Screening_Fail,
    Screening_Success]
    Screening_Entrance;
        :
    /* Service Definition */
        :
    ( Screening_Fail; exit [] Screening_Success; exit )
endprocess
)
```

2.3. Specification Architecture

In general given n services, S_i where $1 \leq i \leq n$, let

S_0 be the basic call model/service process,
DP   be the set of events (detection points) in S_0 (eg. Analyze_Info and Analyzed_Info),
SE_i be the starting event for service S_i where $SE_i \in DP$,
TE_i be a set of terminating event, $(te_x)_i$ for service S_i where $(te_x)_i \in DP$ and $x > 0$,
EE_i[1]   be the entrance event for service S_i,
XE_i be a set of exit events, $(xe_m)_i$ for service S_i where $m > 0$,
$Link_i$ be the linking process for service S_i based on SE_i, TE_i, EE_i and XE_i.

the overall specification architecture expressed in LOTOS is as follows:

```
S0 [DP]
|[DP]|
( Link1 [SE1, TE1, EE1, XE1]
    |[EE1, XE1]|
    S1 [EE1, XE1]
|||   Link2 [SE2, TE2, EE2, XE2]
    |[EE2, XE2]|
    S2 [EE2, XE2]
        :
|||   Linkn [SEn, TEn, EEn, XEn]
    |[EEn, XEn]|
    Sn [EEn, XEn]

    where
        process Link1 [SE1, TE1, EE1, XE1]
            SE1; EE1
            (   (xe1)1; (te1)1; exit
            []  (xe2)1; (te2)1; exit
                :
            []  (xep)1; (tep)1; exit
            )
```

---

[1] Note that only one entrance event is specified in this architecture. It is possible to allow two or more entrance events to be associated with a terminating process, however a resolution mechanism is needed to specify which entrance event should be the starting point if two or more entrance events are active simultaneously.

endprocess

process S_1 [EE_1, XE_1]
  EE_1;
  $\vdots$
  /* Service S_1 Definition */
  $\vdots$
  ($(xe_1)_1$; exit
  [] $(xe_2)_1$; exit
  $\vdots$
  [] $(xe_p)_1$; exit
)
endprocess
$\vdots$

process $Link_n$ [SE_n, TE_n, EE_n, XE_n]
  SE_n; EE_n
  ($(xe_1)_n$; $(te_1)_n$; exit
  [] $(xe_2)_n$; $(te_2)_n$; exit
  $\vdots$
  [] $(xe_q)_n$; $(te_q)_n$; exit
)
endprocess

process S_n [EE_n, XE_n]
  EE_n,;
  $\vdots$
  /* Service S_n Definition */
  $\vdots$
  ($(xe_1)_n$; exit
  [] $(xe_2)_n$; exit
  $\vdots$
  [] $(xe_q)_n$; exit
)
endprocess
)

Note that each of the above services can also be further decomposed such that additional chaining is possible within each process (section 4).

3. Introducing Constraints on Service Specification

3.1. Management of Service Priority and Invocation Precedence

By applying constraints during process composition, it is possible to manage and resolve control interactions involving BCSM and services such as those presented in [4]. It includes situations where two or more services may be invoked at and/or returned to the same synchronisation point (event). For instance, the invocation of two services can be represented as two separate processes shown in Figure 6 below. Note that for simplicity Service_1 and Service_2 represent both the starting and terminating events.

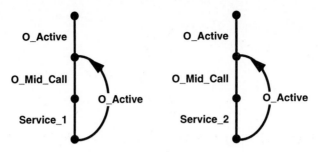

Figure 6. Invocation of two separate services

The composition of Figure 6 and Figure 2 based on synchronisation on events O_Mid_Call and O_Active (as in section 2) is shown in Figure 7. The resulting composition has two possible choices of services (ie. non-determinism).

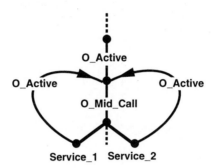

Figure 7. Service composition

The composition of Figures 2 and 6 can be expressed in the following partial LOTOS process, *compose* :

process compose [...]
  BCSM [O_Mid_Call, O_Active, ...]
  | [O_Mid_Call, O_Active] |
  (Service_1_Link [O_Mid_Call, O_Active, Service_1]
      | [Service_1] |
      Service_1_Definition [Service_1]
  ||| *Service_2_Link [O_Mid_Call, O_Active, Service_2]*
      | [Service_2] |
      Service_2_Definition [Service_2]
)

To overcome non-determinism, it is necessary to introduce a separate process that controls the precedence of service invocation, sets service priority and/or disables certain services ie. puts a constrain on the composition. The same synchronisation and composition approach can be used again. For instance a new process can be specified to force the invocation of Service_1 only and to disable Service_2. The new process is shown in Figure 8.

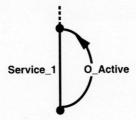

Figure 8. Process to enable Service_1 and disable Service_2.

The resultant composition expressed in LOTOS is

compose [...]
| [Service_1, Service_2] |
(Service_1; exit)

Alternatively, the control may be such that you wish to alternate between the two services (this is an illustrative exercise and it is not meant to represent handling of service interaction). The process may be expressed as follows :

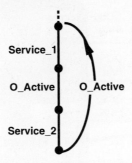

Figure 9. Process to alternate between Service_1 and Service_2.

The resultant composition expressed in LOTOS is

compose [...]
| [Service_1, Service_2] |
(Service_1; Service_2; exit)

If you allow the system to randomly choose between the two services then you may specify the process as follows :

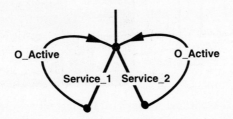

Figure 10. Random selection of Service_1 or Service_2.

The resultant composition expressed in LOTOS is

compose [...]
| [Service_1, Service_2] |
(Service_1; exit [] Service_2; exit)

4. Example

The following example (Figures 11 and 12) illustrates how two separate services can be modelled based on the above approach, and shows the ordering of service invocation can be specified using the same mechanism. The service features considered are Originating Call Screening (OCS) and Abbreviated Dialling (ABD). Note that the SIB concept [5] is used to model these features. Instead of BCSM, the BCP SIB is used to show that the call model may be represented at the SIB level of abstraction. The example only illustrates the control aspect of service modelling and therefore the two PORs (Point of Return): Proceed With New Data and Continue With Existing Data for BCP SIB are simplified to Continue POR.

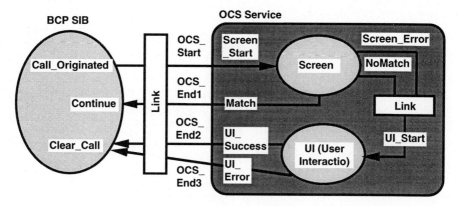

Figure 11. Modelling Originating Call Screening service using SIBs.

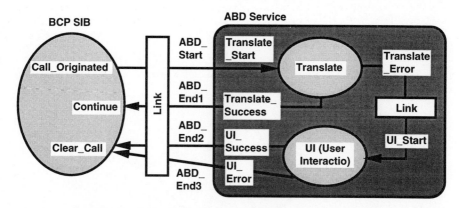

Figure 12. Modelling Abbreviated Dialling service using SIBs.

The partial BCP SIB can be represented as follows (Figure 13):

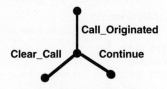

Figure 13. BCP SIB.

The Originating Call Screening and Abbreviated Dialling services are shown in Figures 14 and 15, respectively.

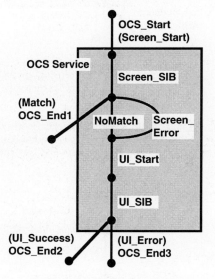

Figure 14. Originating Call Screening (OCS) service. Note that events Screen_Start, Match, UI_Success and UI_Error are relabelled as OCS_Start, OCS_End1, OCS_End2 and OCS_End3 respectively.

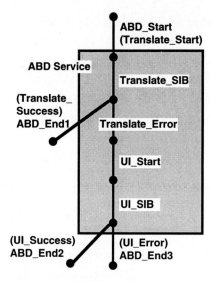

Figure 15. Abbreviated Dialling (ABD) service. Note that events Translate_Start, Translate_Success, UI_Success and UI_Error are relabelled as ABD_Start, ABD_End1, ABD_End2 and ABD_End3 respectively.

The services in Figures 14 and 15 are specified independently based on the BCP SIB. In order to relate the services to the overall call model, two additional processes, OCS_Link and ABD_Link (Figures 16 and 17) are introduced to link OCS service and ABD service to the BCP SIB.

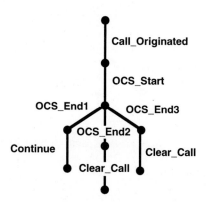

Figure 16. Linking Originating Call Screening service (OCS_Link) to BCP SIB.

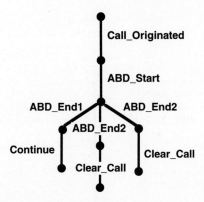

Figure 17. Linking Abbreviated Dialling service (ABD_Link) to BCP SIB.

The compositions of OCS and OCS_Link, and ABD and ABD_Link expressed in LOTOS are as follows:

- (OCS |[OCS_Start, OCS_End1, OCS_End2, OCS_End3]| OCS_Link) and
- (ABD |[ABD_Start, ABD_End1, ABD_End2, ABD_End3]| ABD_Link)

The approach illustrated so far presents a way of integrating services to the call model. In this particular case, the two services can be invoked at the same POI (Point of Invocation): Call_Originated. Assuming it has been decided that ABD service must be invoked prior to OCS service, the following Service Interaction Management (SIM) process expresses the above precedence rule (Figure 18).

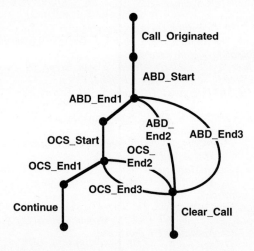

Figure 18. Service Interaction Management (SIM) process.

The overall architecture and interaction between SIBs in the above example can be represented by the following partial LOTOS specification. For simplicity, the behaviours of SIBs are not specified in detail.

BCP [Call_Originated, Continue, Clear_Call, ...]
||| (OCS_Link [Call_Originated, Continue, Clear_Call, OCS_Start, OCS_End1,
            OCS_End2, OCS_End3]
         | [OCS_Start, OCS_End1, OCS_End2, OCS_End3] |
         OCS [OCS_Start, OCS_End1, OCS_End2, OCS_End3])
||| (ABD_Link [Call_Originated, Continue, Clear_Call, ABD_Start, ABD_End1,
            ABD_End2, ABD_End3]
         | [ABD_Start, ABD_End1, ABD_End2, ABD_End3] |
         ABD [ABD_Start, ABD_End1, ABD_End2, ABD_End3])
)
| [OCS_Start, OCS_End1, OCS_End2, OCS_End3, ABD_Start, ABD_End1,
  ABD_End2, ABD_End3] |
SIM [OCS_Start, OCS_End1, OCS_End2, OCS_End3, ABD_Start, ABD_End1,
  ABD_End2, ABD_End3]
where
  process BCP [Call_Originated, Continue, Clear_Call, ...]
      Call_Originated;
      (Continue; ...
      [] Clear_Call; ...)
  endprocess

  process OCS_Link [Call_Originated, Continue, Clear_Call, OCS_Start, OCS_End1,
      OCS_End2, OCS_End3]
      Call_Originated; OCS_Start;
      (OCS_End1; Continue; exit
      [] OCS_End2; Clear_Call; exit
      [] OCS_End3; Clear_Call; exit)
  endprocess

  process OCS [OCS_Start, OCS_End1, OCS_End2, OCS_End3]
      OCS_Definition [OCS_Start, OCS_End1, OCS_End2, OCS_End3]
      where
          process OCS_Definition [Screen_Start, Match, UI_Success, UI_Error]
              Screen_SIB [Screen_Start, Match, NoMatch, Screen_Error]
              | [NoMatch, Screen_Error] |
              (Screen_UI_Link [NoMatch, Screen_Error, UI_Start]
              | [NoMatch, Screen_Error, UI_Start] |
                  UI_SIB [UI_Start, UI_Success, UI_Error])
              where
                  process Screen_SIB [Screen_Start, Match, NoMatch, Screen_Error]
                      Screen_Start; i;
                      /* Screen_SIB_Definition in a similar manner as OCS_Definition*/
                      (Match; exit
                      [] NoMatch; exit
                      [] Screen_Error; exit)
                  endprocess

                  process Screen_UI_Link [NoMatch, Screen_Error, UI_Start]
                      (NoMatch; UI_Start; exit
                      [] Screen_Error; UI_Start; exit)
                  endprocess

                  process UI_SIB [UI_Start, UI_Success, UI_Error]
                      UI_Start; i;
                      /* UI_SIB_Definition in a similar manner as OCS_Definition*/
                      (UI_Success; exit
                      [] UI_Error;exit)
                  endprocess
          endprocess

endprocess

process ABD_Link [Call_Originated, Continue, Clear_Call, ABD_Start, ABD_End1, ABD_End2, ABD_End3]
   Call_Originated; ABD_Start;
   (*ABD_End1; Continue; exit*
   [] ABD_End2; Clear_Call; exit
   [] ABD_End3; Clear_Call; exit)
endprocess

process ABD [ABD_Start, ABD_End1, ABD_End2, ABD_End3]
   ABD_Definition [ABD_Start, ABD_End1, ABD_End2, ABD_End3]
   where
      process ABD_Definition [Translate_Start, Translate_Success, UI_Success, UI_Error]
      Translate_SIB [Translate_Start, Translate_Success, Translate_Error]
      | [Translate_Error] |
      (*Translate_UI_Link [Translate_Error, UI_Start]*
      | [UI_Start] |
         UI_SIB [UI_Start, UI_Success, UI_Error])
      where
         process Translate_SIB [Translate_Start, Translate_Success, Translate_Error]
         Translate_Start; i;
         / Translate_SIB_Definition in a similar manner as ABD_Definition */*
         (*Translate_Success; exit*
         [] Translate_Error; exit)
      endprocess

         process Translate_UI_Link [Translate_Error, UI_Start]
         Translate_Error; UI_Start; exit
      endprocess
   endprocess
endprocess

process SIM [OCS_Start, OCS_End1, OCS_End2, OCS_End3, ABD_Start, ABD_End1, ABD_End2, ABD_End3]
   ABD_Start;
   (*ABD_End1; OCS_Start;*
      (*OCS_End1; exit*
      [] OCS_End2; exit
      [] OCS_End3; exit)
   [] ABD_End2; exit
   [] ABD_End3; exit)
endprocess

5. Conclusion and Future Work

This paper presents a technique based on parallel composition and synchronisation of processes in the process algebra framework to support incremental service specification and to manage service interaction. The approach is formalised using LOTOS and could be used to formally define service interaction. The behaviour chaining process described above requires a service designer to identify and specify link between services/service features so that new services can be defined. Since these links are the only way the processes interact, the analysis of interaction can be focused on these links.

For instance looking at the example in Figures 11 and 12 where the two services are composed, we can identify events that are associated with the links alone (not the internal events in each process) and develop a trace of the link events for each service. The traces of link events for OCS_Service and ABD_Service (Figures 16 and 17 respectively) are (in LOTOS notation):

- Call_Originated; OCS_Start;
  (OCS_End1; Continue [] OCS_End2; Clear_Call [] OCS_End3; Clear_Call)
- Call_Originated; ABD_Start;
  (ABD_End1; Continue [] ABD_End2; Clear_Call [] ABD_End3; Clear_Call)

The traces can than be composed using different synchronisation mechanisms (eg. || or |[..]| operators in LOTOS) and used as the basis for identifying interaction. For example, non-determinism arises as the result of composing the two traces above based on the starting event, Call_Originated (|[Call_Originated]| in LOTOS notation) implies interaction exists. In this case a choice has to be made as to which service is to be invoked. The resultant trace is as follows:

- Call_Originated;
  (OCS_Start; (OCS_End1; Continue [] OCS_End2; Clear_Call [] OCS_End3; Clear_Call)
  [] ABD_Start; (ABD_End1; Continue [] ABD_End2; Clear_Call [] ABD_End3; Clear_Call)
)

The technique presented in this paper provides a rigorous basis for further work on feature interaction definition and analysis. Future work will involve using this approach for other service specifications and interaction analysis.

6. Acknowledgment

This paper reports work performed by the author at the Collaborative Information Technology Institute and the Royal Melbourne Institute of Technology. The author wishes to acknowledge the sponsorship of both organisations and their permission to publish the paper. The author is grateful to Dr. J Cheong of Telecom Australia for valuable discussions and technical assistance, and the anonymous referees for useful suggestions.

7. References

[1] Information Processing Systems - Open Systems Interconnection. LOTOS - A Formal Description Technique based on the Temporal Ordering of Observational Behaviour, ISO 8807, 1989.
[2] C.A.R. Hoare. Communicating Sequential Process, Prentice-Hall, 1985.
[3] CCITT Study Group XI/4. Revised Recommendation Q.1214, Temporary Document XI-12E, Geneva, 9-20 March 1992.
[4] E. Kuisch, R. Janmaat, H. Mulder and I. Keesmaat. A Practical Approach to Service Interactions. IEEE Communications, Vol. 31 No. 8, August 1993.
[5] CCITT Study Group XI and XVIII. Draft Recommendation Q.1203. Intelligent Network - Global Functional Plane Architecture, COM XI-R 108-E/COM XVIII R 72-E, Geneva, 23-25 September 1991.

Use Case Driven Analysis of Feature Interactions

Kristofer KIMBLER, Daniel SØBIRK

Department of Communication Systems,
Lund Institute of Technology,
Box 118, 221 00 Lund, Sweden

Abstract. This paper introduces a user-oriented approach to feature interaction analysis. The presented method is an extension of the Use Case Driven Analysis technique originating from Object Oriented Software Engineering. It aims first at creating the Use-Case Model which describes different possible ways of using the system services, and then at building the Service Usage Model which shows dynamic relations among the services and their features from user's point of view. The actual detection of feature interactions starts with generating possible call scenarios from the Service Usage Model. The sequences of events in the call scenarios are then analysed to detect interaction-prone combinations of the features. The application of the method to IN services, as well as the algorithm used for generating call scenarios and for detecting interaction-prone feature pairs are presented. A prototype tool support for the method is briefly discussed, and a few example of detected feature interaction are given.

1. Introduction

The problem of Feature Interaction (FIN) [1] can be addressed in different phases of the service life-cycle and on different levels of abstraction. There are tendencies to analyse, detect, and solve FINs during the service creation and network implementation phases. The objective of this approach is to make the service implementation "interaction-free", i.e. to eliminate all undesirable FINs before service deployment is done. Analytical and simulation methods are used for this purpose [6,11], together with formal methods [4,6], and formal specification languages like SDL, LOTOS, or Z [3, 5, 7].

On the other hand, there are attempts to detect and solve FINs after the service deployment, i.e. during activation, parametrisation, and execution of the services. The advocates of this approach claim that it is impossible to analyse FINs during service creation, because the number of feature combinations to be analysed grows exponentially with the number of features in the network [2]. The objective is to allow rapid and independent service creation. Therefore, instead of extensive analysis of FINs during service creation some run-time network solutions like Feature Interaction Manager or Negotiating Agent Model [8] are proposed in order to detect and resolve undesirable FINs during service execution.

The feature interaction analysis method presented in this paper falls into the first category. In principle, the method applies the Use Case Driven approach to analyse the services and their features in order to detect interaction-prone combinations of service features [11]. The Use Case Driven Analysis is a widely spread technique in Object Oriented Software Engineering [9,10]. During the analysis an informal requirements definition of services and their features is transformed to a model which can serve as a requirements specification. The model is then used for detecting interaction-prone feature combinations. In that sense, the presented method can be regarded as an integral element of the requirements analysis phase in the service creation process.

In the subsequent chapters we will present the FIN analysis method, illustrated by an example, and the tool support for the FIN analysis.

2. FIN Analysis Method

2.1. Use Case Driven Analysis

As mentioned above, the Use Case Driven Analysis (UCDA) is an Object Oriented technique [9,10] that analyses the system requirements, in our case Intelligent Network, starting from user's point of view. The analysis aims at finding different possible scenarios of system usage, called use cases. In our application of the UCDA, we try to find how the IN services could be used and combined by different kinds of users. During the analysis several system models are built. The most important are the *Use Case Model* (UCM) and the *Service Usage Model* (SUM) [12]. The first one describes actors and use-cases, whereas the second contains the dynamic models of the services. The process of UCDA is shown in fig. 1.

FIGURE 1. Use Case Driven Analysis

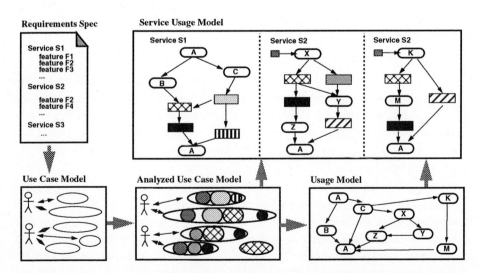

We assume that some form of *Requirements Specification* exist before the analysis is started. In case of IN, it might be a list of services and their features together with the architectural, management, signalling, and capacity requirements. Our focus is, of course, on the functional requirements - the list of services and features. We assume that the services are described as commercial packages consisting of basic functional components - features. The functionality of every feature together with its activation, parametrisation, and invocation procedures are described as well. We also assume that requirements are specified in prose and does not contain any formal descriptions of the services and features. Creation of such precise descriptions is a one of the main objectives of our method. In the following sections we will discuss the process of constructing the UCM and SUM.

2.1.1. Use Case Model

The Use Case Model (UCM) contains actors and use cases. An *actor* is a *role* played by the *user*. In case of IN, we might, for instance, distinguish two basic actors, service user and service subscribers. For instance, a person calling a free-phone number plays the role of service user, whereas the person providing this free-phone service (e.g. shopping information) plays the role of service subscriber. A charge card caller might play the role both of service subscriber and service user at the same time! These two basic kinds of actors can be further specialised to business and private users, business and private subscribers, mobile and stationary business users, etc. We can say that each actor models different needs and different behaviour of potential system users. Actors might be also used to model the roles of non-human users, i.e. other systems or devices interacting with the system in question.

By analysing potential behaviour of each actor we identify different *use cases*. A use case is a specific scenario of the system usage described as a sequence of events and user interactions with the system. In the case of IN, use cases should describe different call scenarios and different ways of using and combining IN services. We can say that the UCM is structured by actors, i.e. for each actor a number of relevant, characteristic use cases should be described. Of course, in practice, most of the use cases will involve more than one actor (e.g. A service subscriber and B service subscriber). Use cases are usually described in plain textual form, however more formal techniques like Message Sequence Charts are also used for this purpose. An example of a use case "a free-phone call made by using virtual card" is given below.

TABLE 1. An use case – calling free-phone from VCC.

	Action
1	User A lifts the receiver.
2	A dials the VCC access code.
3	A enters own account number and PIN code.
4	A selects a language.
5	A dials an abbreviated number.
6	A is connected to a free-phone service.
7	A selects the desired kind of service
8	A waits for an answer.
9	A talks to the operator.
10	A hangs up and terminates the call.

Nobody should expect that the use cases identified and described in the UCM will cover all the possible combination of services and features and all the exceptional situations that might happen during the call processing. This should not even be our goal. Use cases are more like structure means that help us to analyse the functionality of a complex system (IN) looking at it from different angels.

Creation of the UCM might look as a quite difficult task when we develop a completely new kind of system. In this situation we have to use our imagination supported by the problem domain experts [9]. Fortunately for us, many advanced telephony services have been available on the switch-based, ISDN, and even IN platforms for several years. Therefore, the market analysis, and even statistics collected by network operators and service providers could be utilised for constructing a relevant UCM.

Moreover, in case of telecommunication systems we rarely deal with the development of an entirely new system. Instead the typical development project in this area aims at extending or modifying an existing system. IN services tend to be created and deployed one by one, or in small packages. The incremental development of IN allows for reuse of UCMs created earlier, or in other words, to evolve the UCM together with the system itself.

2.1.2. Service Usage Model

The first phase of the analysis results in UCM consisting of actors and their use-cases. In the next step we transform the UCM to the *Service Usage Model* (SUM) which describes the dynamic behaviour of the system services from user's point of view. Similarly to the functional requirements specification, the SUM is structured by services, i.e. each service is modelled by a separate *Service Usage Graph* (SUG). A SUG is a state oriented diagram demonstrating dynamic relations (transitions) among the user states and the functional elements (features) of the services.

By breaking the SUM into a number of smaller SUGs, we make the model more readable. We can also much better handle its complexity, especially the state explosion effect so typical for state-oriented descriptions of telecom systems. It should be stressed that from the conceptual point of view all the SUGs form one, integral model of the system. This and other important aspects of the SUM as well as the process of its creation are explained below.

The process of building the SUM is done in several steps. First, the use cases described in the UCM have to be analysed and elaborated. The informal description of each use case is converted into a sequence of *events* which represent user stimuli (e.g. number dialling), system responses (e.g. busy tone), invocation of service features (e.g. authorisation-verification), and other system actions which can be identified either as basic call processing actions (e.g. number analysis), or complex operations involving more then one feature (e.g number translation with time and origin-dependent routing). The descriptions of the use cases resulting from this analysis are described in an auxiliary model called the *Analysed Use Case Model* (AUCM).

In the next step we continue the use case analysis to identify so called *usage states*, i.e. external system states in which the user can make different decisions, i.e. continue the call in different ways. For instance, being in the *dialling* state, the user may dial a "normal" PSTN number, a free-phone number, or even a service access code. In order to properly identify the usage states, and possible user choices in those states, we have to find and analyse similar sub-sequences of events in different use cases. Such analysis helps to determine how the user choices related to the same usage state differ from one service to another, e.g. what the user can to do in the *connected* state when either VPN or VCC is invoked. Below, an example of analysed use case is presented.

TABLE 2. An analyzed use case – calling free-phone from VCC.

	state	event	feature	comments
1	idle	off-hook A	ACC	network access is controlled
2		dialing tone	ORD-C	ordinary charging parameters are set
3	dialling	VCC access code	SAC	the user dials the VCC service access code
4		account no. + PIN code	AUTV	account no. and PIN code are verified
5			CRL	credit limit on user's account is checked
6			ALT-C	alternative charging parameters are set
7	dialling	language selection code	LGS	the user selects desired language
8	dialling	abbreviated number	SPD	the user uses speed dialing feature
9	analysing	free-phone number		free phone no detected
10			REV-C	reverse charging parameters are set
11		prompt/ selected service	OUP	the user selects the desired service
12			TDC	parameters for routing features are set
13			ODR/TDR	an appropriate route is selected
14			CD	an appropriate destination is selected
15		queue announcement	QUE	call is waiting in a queue
16	ringing	off-hook B		the user is connected to an operator
17	connected	on-hook A		the user hangs on and is disconnected
18			CHRG	the charging procedure is done
19	idle			the call is finished

The next auxiliary model built as a step-up towards the SUM is the *Usage Model* (UM). The Usage Model describes usage states and their transitions. This approach shows many similarities to the modelling technique developed in Statistical Usage Testing [11]. However, our purpose is not to describe the statistical properties of user's behaviour (state transition probabilities), but to capture the functional requirements from user's point of view.

Please notice that some of the user dialogue features like Outgoing User Prompter (OUP) have similar properties to usage states, i.e. they allow the user to make different choices. Though, for the sake of the feature interaction analysis we will not treat them as usage states. Moreover the user dialogue (stimuli/responses) related to these features will be hidden.

Eventually, having both AUCM and UM ready we can create the *Service Usage Model* (SUM). As mentioned above, the SUM consists of a number of *Service Usage Graphs* (SUGs), one SUG for each service, and in that sense its structure corresponds to the functional requirements (see fig. 1). The SUGs are derived from UM by "filling" the Usage Model with the sequences of actions described in the analysed use cases. By doing so we specify what actually happens when the system goes from one usage state to another, and this way we give the semantics to the usage state transitions. The usage states and the system actions (e.g. invocation of service features) form the nodes of the SUGs, whereas the user stimuli and system responses are used to mark the transitions they cause.

While constructing the SUG for a particular service, we do not need to utilize the whole UM. Only a subset of the usage states and their transitions that is relevant for the service is used. We build each SUG separately from the others. It is quite natural that, during this creative process, several internal system states and their transitions might be discovered and incorporated into the SUGs. Also inconsistencies between AUCM, and UM might be discovered. In such cases the process has to be iterated to make appropriate corrections.

Since different services can be combined within one use case (e.g. a PRM number can be called from VPN) we have to express this fact in our SUM. For this purpose we introduce the notion of *common states*. A common state is a usage state (e.g. dialling) or an internal system state (e.g. number analysis) which occurs in more then one SUG. Common states are used to model possible service combinations in an implicit way. Roughly speaking, when "call processing" reaches such a common state, it may continue in any other SUG where this state exists, provided that the service represented by the other SUG has already been invoked during the call, or the state is the entry state of the other service. The name of common states are global in the SUM. Common states will be further discussed below.

Service Usage Model does not really distinguish between IN services and switch-based services. In particular, the POTS is treated as yet another service offered to the user. The user-oriented approach to the service analysis is main rational behind it. The user does not see any BCPM, and in fact does not even care on which platform and how the service is provided (unless the price is higher!). Consistently, we utilise special features and states to represent actions like "ordinary" charging (CHRG) or number analysis. This way the discussed method can be applicable for analysing services of any other network platform, e.g. ISDN.

FIGURE 2. Example of Service Usage Model

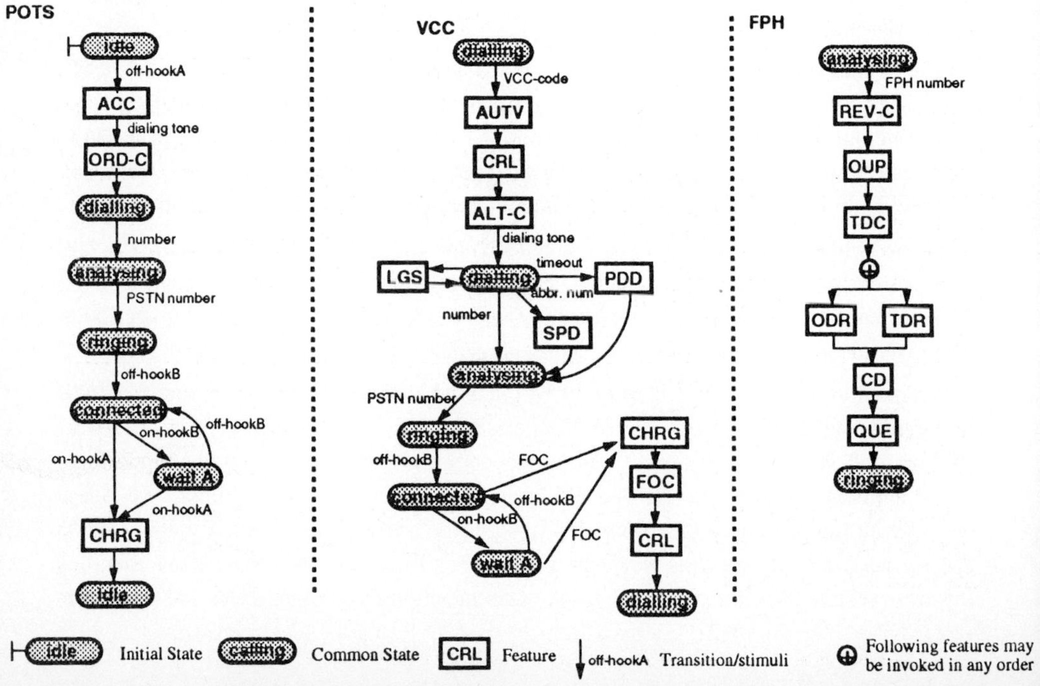

An example of SUM is given in fig. 2 above. The SUM presented above contains just three SUGs for POTS, VCC (Virtual Card Call), and FPH (Free-Phone). The SUGs representing the services are fairly simplified for the purpose of this paper, i.e. the contain only a limited number of features and describe only some basic transitions. There is only one initial state (idle) located in the POTS service. All the possible complete call scenarios (use cases) start and end in this state.

2.1.3. Generation of Call Scenarios

The AUCM and the SUM must be consistent in order to assure the completeness and traceability of the model transformation. This means that for each use case described in AUCM, it should be possible to generate its sequence of events from SUM. In other words, the *call scenarios*, i.e. flows of events that can be generated from SUM, should at least cover all the use cases contained in AUCM. Of course, the number of call scenarios that can be generated from SUM is usually much higher than the number of use cases in AUCM.

To illustrate this point, let's consider again our system containing three services POTS, VCC, and FPH as presented in fig. 2 in the previous section. Starting from the initial state (idle), and traversing the SUGs we are able to re-generate our previous example of a use case "a free-phone call made by using virtual card". Please notice, that the flow of events in this scenario "traverse" all the three SUGs. The scenario is described in the table 3 below.

TABLE 3. Example of call scenario – Calling FPH from VCC

no	POTS	VCC	FPH	Comments
1	idle			off hook
2	ACC			network access is checked
3	ORD-C			normal charging is set up
4	dialling	dialing		VCC–code is dialled
5		AUTV		access and PIN code is validated
6		CRL		credit limit is checked
7		ALT-C		VCC charging is set up
8		dialling		digits are accepted
9		LGS		language select code is dialled and language entered
10		dialling		digits are accepted
11		SPD		abbreviated number
12		analysing	analysing	switch analysis of destination number
13			REV-C	free-phone charging is set up
14			OUP	Prompt in previously chosen language
15			TDC	changes call distribution policy
16			ODR/TDR	select the route
17			CD	selects the destination
18			QUE	queuing
19			ringing	phone is ringing
20	connected		connected	talking
21	CHRG			call is charged
22	idle			

Call scenarios generated from the SUM are used for detecting feature interactions. In that sense, SUM is the central element of the whole method. The actual detection method is described and discussed in section 3.2.

3. Analysis of Feature Interactions

3.1. Basic Assumptions

The method described in this paper is constrained to detection of feature interactions occurring within the scope of one call instance. However, it can be easily extended to detect interactions occurring between features used in different call instances. Interactions caused by deployment, activation, or parametrisation of the services and features are not addressed here. This is simplification of the problem, but still as we know the majority of the interactions occur when different features are used within the same call instance.

The method is based on the simple observation that features are not used as stand-alone entities. Instead, they are packaged into services. If two services cannot be used within the same call, we might expect that no interaction between their features can occur. The method assume that two features can interact if and only if we are able to generate a *call scenario* (from SUM) in which both features occur. Of course, there might be many call scenarios for a given feature interaction case. This assumption holds equally for features belonging to the same service or to two different services. As mentioned above we always analyse feature interactions in service context even when the features belong to the same service.

3.2. Detection of Feature Interactions

The detection of feature interactions starts when the Service Usage Model is completed. The SUM is utilised to determine which combinations of services are possible within one call, and which combinations, or rather sequences of features, can occur within each of the services. Combining these data, we can learn which sequences of features can occur within call scenarios generated from SUM, and thus we are able to point out the service contexts for interaction-prone pairs of features.

In the current version of the method we assume that a combination of two service features is interaction-prone when the features access of modify the same service or call specific data. This is a very simple, not to say rough, approximation. However, it is hard to find an example of feature interaction where the involved features do not use the same data entities. Further development of the method should bring more sophisticated criteria for selecting interaction-prone service pairs.

For every feature a list of the call specific and service specific data entities used by the feature must be provided. Data entities are represent simply by names. Before the actual analysis starts, all the pair-wise combination of the features occurring in the SUGs are checked in order to pre-select the pairs of features which use the same data entities. These interaction-prone pairs will then be searched for during the analysis of call scenarios.

The algorithm of finding service context for a given interaction-prone pair of features will be described in the next chapter. Here we just should say that the interaction-prone pairs of features together with their service contexts are considered as the output of the method and the input to further manual analysis.

4. Tool Support for FIN Analysis

4.1. Objectives of the Tool Support

The aim of the tool support is to make it possible to use the FIN analysis method on a large scale. The unwieldy number of possible feature combinations makes a thorough manual analysis cumbersome, not to say impossible, especially if the number of services and features will continuously grow. Hence, a tool that can automatically point out all the interaction-prone combinations of features together with their service contexts is very desirable. Still, however, the manual analysis of the interaction-prone combinations of features is necessary.

The tool support for the method is under development at the Department of Communication Systems, Lund Institute of Technology. Currently the tool is on the prototype stage and supports only the generation of call scenarios and the detection of interaction-prone feature combinations. This prototype tool is described in the next section. The complete tool should provide the following functionality:

- Graphical support for specification of features and use cases.
- Graphical support for creation of (Service) Usage Graphs.
- Automatic generation of call scenarios from SUM.
- Automatic detection of interaction-prone feature combinations.

4.2. Prototype Tool Support

As mentioned above, the current version of the tool provides only automatic detection of interaction-prone feature combinations. The tool implements in principle the FIN detection algorithm described in section 3.2. The tool requires the following input:

- Formal description of services - SUGs (currently in the script format only). A SUG provides information about dynamic relation between services and features.
- Description of features (also in the script format) including the service and call specific data accessed or modified by each feature.

The detection algorithm has three phases. First, the tool analyses all pair-wise feature combinations in the service-free context and makes a list of the pairs that are interaction-prone (i.e. share the same data). Then, by exploring the SUGs, the tool generates all possible sequences of features and states for each SUG. These sequences are further analysed to detect all potential combinations of services (according to the SUM), and all intra-service feature interactions, i.e. interaction-prone pairs of features occurring within one service. The tool also checks the consistency of the SUGs in terms of correctness of state transitions.

For every common state the set of features that can be invoked after this state is determined as well. Since all the SUGs are taken under consideration, such a set contains features belonging to different services. This is quite natural, because the state is *common* to several services. This information is used in phase three for detecting inter-service feature interactions, i.e. cases when two interaction-prone features are invoked in different services.

Afterwards, the state exploration of SUGs is repeated, this time to detect inter-service feature interactions. For each SUG all possible paths are generated once more. When a path "reaches" a common state, then the features that occur in the path (so far) are matched against the set of features that can follow this state. As we remember the latter was built in phase two of the algorithm, and contains also features from other services. This way inter-service interactions are discovered.

The list of interaction-prone pairs of features detected in phase two and three is the output from the tool. For each pair a list of service combinations in which the pair occurs is provided. Additionally the tool produces a list of potential combinations of services. All this data is then passed to further manual analysis which determine, for each generated pair of features, whether there is an undesired interaction or not.

The SUM provides naturally for generation of complete call scenarios, i.e. sequences of events that begin and end in the initial state (idle). We might use this possibility, and analyse the sequences of features in each and every scenario that can be generated. This, however, will dramatically increase the complexity of the method, which in turn will make it impossible to provide any reasonable tool support. Complete call scenarios could be generated to provide samples giving detailed information about the context in which an interaction-prone pair of features has been detected.

4.3. Example of FIN Detection

Using the simple SUM presented in fig. 2, the prototype tool managed to produce a number of interaction-prone pairs of features. Below, three examples of the detected interaction cases are presented. Please notice that all three cases can be found in the sample call scenario shown in table 3. The explanations provided for the cases demonstrate the reasoning that might be used during manual verification of the interaction-prone pairs.

1. **VCC.LGS - FPH.OUP.** Outgoing User Prompter, OUP, in FPH (step 14 in table 3) prompts the user with some kind of recorded announcement. This announcement should be in the language the user previously selected with LGS in VCC (step 9). If OUP plays a message in a language different from the language selected in LGS, the user will consider this as an inconsistent behaviour of the system. In this case, the interaction is detected because OUP uses a language identifier (call data) that is modified by LGS.

2. **FPH.TDC - FPH.CD.** Time Dependent Control, TDC, (step 15) may affect the parameters (service specific data) of Call Distribution, CD, QUE, etc. in the same service FPH. This interaction is of the same kind as the previous one.

3. **VCC.CRL - FPH.REV-C.** The Credit Limit, CRL, in VCC (step 6) disables the service if user's credit limit has been exceeded. On the other hand, the user should be allowed to make calls to free-phone numbers without being charged at all, even when the credit limit is empty. CRL interacts with Reverse Charging, REV-C, in FPH (step 13) because both features access charging data. Please notice that the described case is not directly related to charging; it is more like a logical interaction between the two services, VCC and FPH. This, by the way, is a difficult interaction case to solve.

5. Summary and Conclusions

The FIN detection method presented in the paper is based in the Use Case Driven Analysis of system requirements. During the analysis the informal description of the services and features is transformed to the Use Case Model and then to the Service Usage Model. The first model describes how different kinds of users can combine system services in different use cases. The second model describes formally the dynamic dependencies between the services and their features by means of Service Usage Graphs. From the Service Usage Model all the possible call scenarios can be generated. The models are consistent if all the use cases from UCM can be re-generated from SUM as call scenarios. The SUM is automatically analysed in order to detect interaction-prone feature pairs, i.e. combinations of features accessing the same data, which can occur in one call scenario. Such interaction-prone cases are then manually analysed to determine if any undesirable interaction might occur.

The Use Case Driven Analysis shows many similarities to the approach taken by experimental physics. Having a hypothesis, we try to prove it by experiments. When sufficient experimental material proving the hypothesis has been collected, we analyse it and build a formal (mathematical) model of our hypothesis. Now we try to repeat (simulate, calculate) our experiments in this formal model. If we are able to obtain the same results as in reality we might assume that the hypothesis is a true fact.

Formalisation of the service analysis process, and formalisation of the service description allow for constructing a tool supporting the detection of feature interactions. In fact, the construction of such a tool is relatively easy, so is the detection method. More effort is necessary to create a tool for the Use Case Driven Analysis, especially the graphical creation of Service Usage Graphs. The method together with the tool will give a great support for service creation which will simplify detection and resolution of undesirable interactions on the early stages of the service life-cycle.

6. References

[1] T. F. Bowen et al. *Views on the Feature Interaction Problem*. Bellcore, USA, September 1990.
[2] T. F. Bowen et al. *The Feature Interaction Problem in Telecommunication Systems*. Bellcore, USA, March 1989.
[3] R. Boumezbeur. *Specifying Telephone Systems in LOTOS and the Feature Interaction Problem*. Canada.
[4] W. Bouma, H. Zuidweg. *Towards Formal Analysis of Feature Interactions*. PTT Research, the Netherlands, 1992.
[5] A. Lee, *Formal Specification - a Key to Service Interaction Analysis*, University of Queensland, 1992
[6] E. Kuisch, R. Janmaat, H. Mulder and I. Keesmaat. *A Practical Approach towards Service Interactions*. PTT Research, the Netherlands, September 1992.
[7] Y. Inoue, K. Takami and T Ohta, *Method for Supporting Detection and Elimination of Feature Interaction in a Telecommunication System*. ATR Communication Systems Research Laboratories, Japan.
[8] N. Griffeth and H. Velthuijsen, *The Negotiating Agent Model for Rapid Feature Development*, Bellcore, USA,
[9] I. Jacobson et al., *Object-Oriented Software Engineering, A Use Case Driven Approach*, Addison-Wesley, 1992.
[10] J. Rumbaugh et al., *Object-Oriented Modelling and Design*, Prentice Hall, 1991
[11] H. Mills et al., Cleanroom Software Engineering, IEEE Software, September 1987
[12] K. Kimbler, E. Kuisch and J. Muller, *Service Interactions among Pan-European Services*, FI Workshop, 1994.
[13] K. Kimbler, *Formalisation of Use Case Driven Analysis, Licentiate Thesis - draft*, Lund Institute of Tech., 1994.

Interaction detection, a logical approach*

Anders Gammelgaard & Jens E. Kristensen
Tele Danmark Research, Lyngsø Allé 2, DK-2970 Hørsholm, Denmark
email:{ag,jek}@tdr.dk

Abstract. We propose to let feature specifications be restrictions to the class of deterministic labelled transition systems. This allows a formal definition of interaction: Two feature specifications interact if each specification has a realization by such a transition system whereas the joint specification has not.

A feature specification consists of two parts: network properties and declarative transition rules. A network property is a condition that constrains all states. A declarative transition rule consists of a precondition, a trigger event, and a postcondition; if a state satisfies the precondition, then it must be the origin of a transition labelled with the trigger event and resulting in a state that satisfies the postcondition.

A simple logic is used to express all conditions on states. The use of logic enables us to model an evolving network since it can describe the relation between old and new network concepts.

1 Introduction

Consider a typical presentation of an interaction, for instance the following between Call Forwarding Unconditional, CFU, and Terminating Call Screening, TCS:

> Suppose B has CFU with calls forwarded to C, and C has TCS with A on his screening list. Then A calls B. Now, the CFU specification says that A will actually initiate a call to C, whereas the TCS specification requires calls from A to be screened. These two requirements are obviously in conflict. Consequently there is an interaction.

We can identify certain key elements in such descriptions:

1. First the current state of affairs is sketched. In the concrete case it is a situation where A and C are idle, B is forwarding incoming calls to C, and C is screening calls from A.

2. Next a trigger event is presented—here a call from A to B. This event triggers actions in one or more of the involved features.

---
*This work was partially supported by the SCORE project. It represents the view of the authors.

3. Finally analysis of the feature descriptions shows that there are conflicting requirements to the resulting state. Concretely, CFU requires that the call event result in a state where A is calling C, which is in conflict with the static TCS requirement that A never succeed in calling C.

In the present paper we demonstrate how such arguments can be made fully formal by taking features to be restrictions to a class of potential network realizations. As candidate realizations we take the class of deterministic labelled transition systems. A label in such a transition system will correspond to a trigger event which causes a transition from one state to another state.

In the above example TCS is described by a static constraint: it must be impossible to reach a state where A is calling C. In contrast CFU is described by a dynamic constraint: if A dials B, then A must actually end up calling C. Full feature specifications consist of both static and dynamic constraints.

Static constraints are formalized by *network properties*. All states in a realization must satisfy a given network property.

Dynamic constraints are formalized by *declarative transition rules*. Such a rule consists of a precondition, a trigger event, and a postcondition. If a state satisfies the precondition, then this state must be origin of a transition labelled by the trigger event, and the resulting state of the transition must satisfy the postcondition. Notice that the assumption of determinism forces the resulting state to satisfy the postcondition of *all* rules which have the same trigger event as the transition and a precondition which applies to the originating state of the transition.

Network properties and the pre- and postconditions of declarative transition rules are formulated in a simple logic which is a restriction of ordinary first order logic.

In the concrete interaction example we focus on two declarative transition rules for CFU and one network property for TCS:

$$idle(A) \xrightarrow{\mathit{off\_hook}(A)} ready(A) \qquad (1)$$

$$CFU_{B \mapsto C} \wedge ready(A) \wedge idle(C) \xrightarrow{dial(A,B)} calling(A, C) \qquad (2)$$

$$TCS_{C:A} \Rightarrow \neg calling(A, C) \qquad (3)$$

The intended interpretation of the symbols should be clear. For instance $calling(A, C)$ means that the telephone is ringing at C while A is waiting for C to do an *off_hook*. Notice that a call really consists of two events, an *off_hook* and a *dial*.

From these constraints it is fairly easy to show that the joint specification of CFU and TCS is unrealizable. For assume the opposite and start from a state where A and C are idle, in a realization where the two features are active in all states. Then, basically using 1 and 2, we derive that there must be a state satisfying $calling(A, C)$. But 3 explicitly forbids the existence of such a state.

It should be noticed that feature descriptions always depend on some basic service. E.g. property 3 only makes sense provided there already exists a basic service which defines the property *calling* and specifies a minimum number of useful transitions. We will take a feature specification to be an extension and/or modification of such a basic service specification. For instance, rule 1 is really a rule from the specification of the basic service, and thus it really belongs to the specifications of both CFU and TCS.

There already exist many formal methods for detecting feature interaction [1, 2, 5, 6, 7, 9, 10, 12, 13]. Two methods are particularly relevant.

The detection and resolution method developed by Ohta et al. [9, 12] has been our major source of inspiration. Their State Transition Rules closely resemble our declarative transition rules. But there are also important differences between the two methods.

Conditions on states are formulated differently. Ohta et al. use sets of so-called state primitives whereas we use predicates. The use of predicates enables us: to carry out standard logical deductions, to support these deductions by explicit network properties, and to cope with evolving network descriptions through logical relations between existing and new concepts. None of these possibilities seem to be directly present with the type of conditions used in State Transition Rules.

The detection conditions also differ. One condition used by Ohta et al. is non-determinacy in selection of rules: If two rules are both enabled in a global state and a certain priority rule does not apply, then there is an interaction. In our method we instead take both rules simultaneously and only detect interaction if the requirements on the resulting or some following state turn out to in conflict. This indicates that the method of Ohta et al. is suited for detection of *interactions* while our method rather deals with the more narrow concept of *interference* [11] or *policy interaction* [8], i.e. with interactions which are observable and undesired for the users.

The real time transition model described by Cameron and Lin in [5] has influenced our method in one important respect. Cameron and Lin describe various enabling and selection policies for transitions in a system, and they argue that a "select-all" selection policy facilitates detection of interactions. The select-all policy corresponds to our requirement that the resulting state of any transition must satisfy *all* declarative transition rules which apply to the transition.

Cameron and Lin use update-inconsistency as the condition for detecting interactions: If, using the select-all policy, two features force some variable to be updated more than once in some time step, then there is an interaction. This condition is also closely related with our detection condition.

The two methods seem to be used quite differently, however. Cameron and Lin use a single variable for recording active features. This gives update-inconsistency as soon as two features start up at the same event, even if the features are not in real conflict. Again this indicates that they aim at detecting interactions while we merely try to detect interference and policy interaction.

The paper is structured as follows: First we discuss feature specifications, we describe how they are formalized and what their intended models are. This paves the way for defining special variants of realizability and interaction. Next we specify a simple core service and extend this into four full feature specifications. Then we outline the formal detection of five interactions. Finally we consider future directions of research.

2 Service versus Feature

Usually features are conceived as extensions and/or modifications to an underlying core service. But since two features may change this service in different directions, we have to use the complete feature specifications—including the modified service—when detecting interactions.

On the other hand the conception of features as modifications suggest a compact way of introducing features: Instead of giving a complete specification each time, we first introduce the core service once and for all, and then just provide the new and the changed rules and properties for each new feature.

In the following we will use the word feature to denote also the specification of a feature, including the description of its underlying core service.

3 Formal specifications

A simple logic is used for expressing network properties and declarative transition rules. It is a restriction of traditional first-order logic as it has no variables, no quantification, and no function symbols. We presuppose:

- A set of *constants* $\mathcal{S} = \{A, B, C, \ldots\}$. Intuitively a constant denotes a subscriber.

- A set $\mathcal{P}_1 = \{dial\_tone, calling, busy\_tone, idle, \ldots\}$ of ranked *state predicate symbols*. These predicate symbols are used to describe the current "state of affairs" in the network.

- A set $\mathcal{P}_2 = \{CW, CFB, TCS, \ldots\}$ of ranked *activation predicate symbols*. They are used to express whether a feature is activated or not; e.g. $TCS(A, B)$ expresses that A has terminating call screening activated, with B on his screening list. We assume a one-to-one correspondence between features and activation predicate symbols.

We assume that $\mathcal{S}$, $\mathcal{P}_1$, and $\mathcal{P}_2$ are disjoint and we denote the union of $\mathcal{P}_1$ and $\mathcal{P}_2$ by $\mathcal{P}$. For pragmatic reasons we also assume that each of $\mathcal{S}$, $\mathcal{P}_1$, and $\mathcal{P}_2$ is reasonably large. This is not a formal assumption, however. Formally the sets may be infinite, finite, or even empty.

Atomic predicates are constructed by applying elements in $\mathcal{P}$ to the right number of arguments in $\mathcal{S}$. *Formulas* are built in the standard way using atomic predicates and propositional logical operators like $\wedge$, $\Rightarrow$, $\neg$, and *true*.

A *network property* is just a formula.

For the description of trigger events we furthermore presuppose:

- A set $\mathcal{T} = \{dial, off\_hook, on\_hook, flash\_hook, \ldots\}$ of ranked *trigger event symbols*.

Trigger events are constructed from $\mathcal{T}$ and $\mathcal{S}$ in the same way as atomic predicates.

A *declarative transition rule* has form:

$$pre \xrightarrow{t} post$$

where *pre* and *post* are formulas and t is a trigger event.

An *axiom* is a network property or a declarative transition rule. A *specification* is a set of axioms.

Schemata We will use schemata to specify sets of axioms.

A *network property schema* is the same as a network property except that we allow meta-variables $x, y, z, \ldots$ to be used in place of constants. A network property schema is an abbreviation for the set of network properties that can be obtained by replacing meta-variables with all possible combinations of constants.

The definition and interpretation of a *declarative transition rule schema* is similar.

In both kinds of schemata we allow side-conditions on the meta-variables. Side-conditions constrain the permitted replacements for the meta-variables. Rule 12 in section 7 is a typical example.

In the following we will not bother to distinguish between a network property and a network property schema or between a declarative transition rule and a declarative transition rule schema.

4 Models

Models of specifications are based on deterministic labelled transition systems. A *labelled transition system* is a tuple $(St, L, \rightarrow)$ where:

- St is a set of states,

- L is a set of labels, and

- $\rightarrow \subseteq St \times L \times St$ is a set of labelled transitions.

As usual we write $s \xrightarrow{l} s'$ in case $\rightarrow$ contains (s, l, s'). A labelled transition system is *deterministic* when:

$$\text{for all } s, s', s'', l, \text{ if } s \xrightarrow{l} s' \text{ and } s \xrightarrow{l} s'', \text{ then } s' = s''$$

Notice that we have not imposed internal structure on neither states nor labels. We will work with an external view instead: A *satisfaction relation* is a relation $\models$ between states in St and formulas of the logic which was introduced in the previous section. This relation must obey all standard relations like $s \models true$, if $s \models f_1$ and $s \models f_2$ then $s \models f_1 \wedge f_2$, etc.

An *interpretation structure* is a pair $((St, L, \rightarrow), \models)$ where $(St, L, \rightarrow)$ is a deterministic labelled transition system and $\models$ is a satisfaction relation. The interpretation structure *satisfies* a network property f whenever:

$$\text{for all } s \in St, \; s \models f$$

And it *satisfies* a declarative transition rule $pre \xrightarrow{t} post$ whenever:

$$\text{for all } s \in St, \text{ if } s \models pre, \text{ then there exists } s' \text{ such that } s \xrightarrow{t} s' \text{ and } s' \models post$$

An interpretation structure is a *model* of a specification if it satisfies all the axioms of the specification together with the *initial state* condition: There exists a state $s_0 \in St$ which satisfies $idle(x)$ for all substitutions of x by a constant from $\mathcal{S}$.

5 Realizability and Interaction

Let F be a feature. An atomic predicate built from the activation predicate symbol corresponding to F is called an F *activation predicate*.

A set of activation predicates is called an *activation record*. Intuitively an activation record is used to restrict the set of interpretation structures to those in which all states satisfy the predicates in the record and no other activation predicates; i.e. an activation record completely describes which feature instances are active and which are not.

A pair (F, AR), where F is a feature specification and AR is an activation record, is *realizable* if F has a model in which all states satisfy exactly those activation predicates which occur in AR.

A feature F is *operate-alone* realizable whenever $(F, \{p\})$ is realizable for all F activation predicates p. Intuitively F is operate-alone realizable if it is possible to implement a network containing just a single active instance of F.

A feature F is *operate-freely* realizable whenever (F, AR) is realizable for all sets AR of F activation predicates. Intuitively F is operate-freely realizable if it is possible to implement a network containing an arbitrary set of active instances of F.

Two features F_1 and F_2 are *cooperate-alone* realizable whenever $(F_1 \cup F_2, \{p, q\})$ is realizable for all F_1 activation predicates p and all F_2 activation predicates q. Intuitively F_1 and F_2 are cooperate-alone realizable if it is possible to implement a network containing just one active instance of F_1 and one active instance of F_2.

Two features F_1 and F_2 are *cooperate-freely* realizable whenever $(F_1 \cup F_2, AR)$ is realizable for all sets AR of F_1 and F_2 activation predicates. Intuitively F_1 and F_2 are cooperate-freely realizable if it is possible to implement a network containing an arbitrary set of active instances of F_1 and F_2.

We will say that a feature is *interacting with itself* if it is operate-alone realizable but not operate-freely realizable.

We will say that two features F_1 and F_2 are *interacting* if each feature is operate-alone realizable but F_1 and F_2 are not cooperate-alone realizable. Notice that this is a relatively narrow definition. Another possible definition would be to take F_1 and F_2 to be interacting if each feature is operate-freely realizable but F_1 and F_2 are not cooperate-freely realizable. The latter definition can only differ from the former in complex cases where one of the features has two or more active instances. Consequently we have chosen to stick with the former and more simple definition.

6 The Core Service

As core service we use a simple Basic Call service BC.

We will have to assume certain capabilities of telephones. Also we will formulate certain natural limitations on telephones.

From a user's viewpoint a telephone has several different states. We formalize these states as basic predicates parameterized by the owner of the phone:

- $idle(x)$: the telephone has the receiver on hook and is silent.

- $dial\_tone(x)$: the telephone emits a dial tone and thus indicates that x can dial a number.

- $busy\_tone(x)$: the telephone emits a busy tone which indicates a failed call attempt or a disconnected line.

- $ringing(x)$: the telephone is ringing, with the receiver on hook.

- $ring\_back\_tone(x)$: this indicates that a ring back tone can be heard at the caller.

- $engaged(x)$: the telephone can be used for communication with a remote destination.

These states are all mutually exclusive. This is formulated by requiring that at most one of the predicates may be true. We use a syntactic shorthand to formalize this network property, taking *false* to be the numerical value 0 and *true* to be the numerical value 1:

$$\sum \{idle(x), dial\_tone(x), busy\_tone(x), \\ ringing(x), ring\_back\_tone(x), engaged(x)\} \leq 1 \quad (4)$$

Since new features may introduce new telephone capabilities, it is too strong to require that the sum *equals* one.

A basic call proceeds through a number of phases. We will now identify and formalize the different phases and subsequently we use the phases to formulate transition rules and additional network properties.

In contrast to the predicates put forward for telephones, the new predicates for the phases of a basic call will not be directly perceivable by a user. They instead define certain internal states of the network.

- *ready*(x): the phone at x is ready for accepting digits.

- *calling*(x,y): this is the phase where the phone is ringing at y and x is waiting for y to accept the call.

- *rejecting*(x): this phase is entered if y happens to be busy when x calls y, or when y puts down the receiver at the end of a conversation with x.

- *path*(x,y): x and y can communicate.

These internal predicates are tied up with the externally observable predicates through a number of network properties:

$$ready(x) \Rightarrow dial\_tone(x) \quad (5)$$

$$calling(x,y) \Rightarrow ringing(y) \land ring\_back\_tone(x) \quad (6)$$

$$rejecting(x) \Rightarrow busy\_tone(x) \quad (7)$$

$$path(x,y) \Rightarrow engaged(x) \land engaged(y) \quad (8)$$

Additional network properties express relationships among the internal predicates. Since we take communication paths to be symmetric (not taking e.g. billing into account), we have the property

$$path(x,y) \Leftrightarrow path(y,x) \quad (9)$$

It is also convenient to use predicates as abbreviations. We will introduce one so far. It just says that a telephone is busy when it is not idle:

$$busy(x) \Leftrightarrow \neg idle(x) \quad (10)$$

The declarative transition rules for BC now describe how to go through the different phases of a Basic Call:

$$idle(x) \xrightarrow{off\_hook(x)} ready(x) \quad (11)$$

$$ready(x) \land idle(y) \xrightarrow{dial(x,y)} calling(x,y) \quad \text{if } x \neq y \quad (12)$$

$$ready(x) \land busy(y) \xrightarrow{dial(x,y)} rejecting(x) \quad (13)$$

$$rejecting(x) \xrightarrow{on\_hook(x)} idle(x) \qquad (14)$$

$$calling(x,y) \xrightarrow{off\_hook(y)} path(x,y) \qquad (15)$$

$$calling(x,y) \xrightarrow{on\_hook(x)} idle(x) \wedge idle(y) \qquad (16)$$

$$path(x,y) \xrightarrow{on\_hook(x)} idle(x) \wedge rejecting(y) \qquad (17)$$

There are also transition rules of a very different kind. These rules specify that certain events do not influence the state of other parts of the system (such rules are usually called frame axioms). For instance we need to say that if there is a path between x and y, then z cannot change this state of affairs by creating an *off_hook* event:

$$path(x,y) \xrightarrow{off\_hook(z)} path(x,y) \quad \text{if } x \neq z \wedge y \neq z \qquad (18)$$

Since there are many rules of this type, we introduce syntactic shorthands. Predicates which should be kept constant may be put in front of a colon preceding the rule. For instance $C \wedge P \xrightarrow{a} C \wedge Q$ is abbreviated to $C : P \xrightarrow{a} Q$. And instead of the $n*m$ rules $C_i : P \xrightarrow{a_j} Q$ for $i = 1,\ldots,n$ and $j = 1,\ldots,m$ we just write a single rule of the form $C_1,\ldots,C_n : P \xrightarrow{a_1,\ldots,a_m} Q$.

Two such syntactic shorthand rules can express that after x has initiated a call, x has full control over the call, except—possibly—for the callee y.

$$ready(x), rejecting(x) : true \xrightarrow{off\_hook(z),dial(z,w),on\_hook(z)} true \quad \text{if } z \neq x \qquad (19)$$

$$calling(x,y), path(x,y) : true \xrightarrow{off\_hook(z),dial(z,w),on\_hook(z)} true \qquad (20)$$
$$\text{if } z \neq x \wedge z \neq y$$

Two other rules say that if x is idle, y can only change this by calling x:

$$idle(x) : true \xrightarrow{off\_hook(y),on\_hook(y)} true \quad \text{if } x \neq y \qquad (21)$$

$$idle(x) : true \xrightarrow{dial(y,z)} true \quad \text{if } \neq (x,y,z) \qquad (22)$$

In the last of these rules we have used the expression $\neq (x,y,z)$ which just means that x, y, and z are pairwise mutually different.

From the axioms which constitute the specification of BC it is possible to derive more network properties and declarative transition rules. For instance 4 and 8 allow us to conclude:

$$path(x,y) \Rightarrow \neg rejecting(y)$$

and 11, 5, and 4 allow us to conclude:

$$idle(x) \xrightarrow{off\_hook(x)} \neg idle(x)$$

As will follow later (section 8) such deductions will be used to detect interactions.

7 Adding Features

We now add features on top of the basic call service. This will provide the specifications which will be used to detect interactions. Furthermore we illustrate that detection of operate-alone unrealizability may be used constructively in the development of new features.

7.1 Terminating Call Screening

The first feature we discuss is Terminating Screen Calling. This feature allows a subscriber to screen incoming calls according to a screening list.

That the screening list of x contains y is formalized using a binary predicate symbol TCS. For convenience we write $TCS_{x:y}$ instead of $TCS(x, y)$.

TCS is naturally conceived of as a restriction to the underlying Basic Call service: If $TCS_{x:y}$ holds, the intention is that x should never receive calls from y. Leaving the case $x = y$ open, this intention is completely formalized by the network property:

$$TCS_{x:y} \Rightarrow \neg calling(y, x) \quad \text{if } x \neq y \qquad (23)$$

One may now analyse whether TCS, including the axioms of BC, is operate-alone realizable and whether it is operate-freely realizable. It is easy to see that neither is the case as the following argument shows.

Assume that $TCS_{B:A}$ holds and that A and B are idle. Then A does an off hook and dials the number of B. These two events correspond to the trigger events of rule 11 and rule 12 respectively. Using the pre- and postconditions of these two rules we now get that $calling(A, B)$ must hold in the final state. But this is an outright contradiction with the new network property 23 for TCS.

It is really no surprise that the new TCS property is in conflict with the axioms of BC since TCS is a feature which constrains the use of BC. The detected conflict just shows that free use of BC may collide with the restriction imposed by TCS.

The conflict may be resolved by giving TCS priority. This is done by substituting rule 12 with a weaker rule:

$$\neg TCS_{y:x} : ready(x) \wedge idle(y) \xrightarrow{dial(x,y)} calling(x, y) \quad \text{if } x \neq y \qquad (24)$$

The strengthening of the precondition essentially makes the effect of a trigger event unspecified in a situation where it was formerly specified: Previously the effect of $dial(x, y)$ was specified in all states satisfying $ready(x) \wedge idle(y)$, but now it is only specified for those states which furthermore satisfy $\neg TCS_{y:x}$.

As a general methodological rule we propose:

(†)   *Introduction of a new feature may not cause formerly specified effects to become unspecified.*

In the particular case of TCS we thus have to specify the effect of calling a TCS-subscriber who has the caller on his screening list. One suggestion is that the caller should receive a busy tone:

$$TCS_{y:x} : ready(x) \wedge idle(y) \xrightarrow{dial(x,y)} rejecting(x) \qquad (25)$$

The complete specification of TCS then consists of the new axioms 23 and 25 together with all the old BC axioms except for rule 12 which must be substituted with rule 24. To sum up, TCS consists of the following axioms:

BC axioms, old and modified	new TCS axioms
4–11, 13–22, 24	23, 25

In the following we will take for granted that this specification of TCS is both operate-alone and operate-freely realizable. We will provide no formal argument for this claim.

7.2 Call Forwarding Unconditional

The CFU feature allows a subscriber to forward all incoming calls to someone else. That y is forwarding all incoming calls to z is written $CFU_{y \mapsto z}$.

For simplicity we will take CFU to be the feature which always forwards calls to the destination, even if the destination has also activated CFU. In other words, in this paper CFU is the feature you and your colleague would use if you wanted to change office for a while.

The formalization of CFU introduces a new predicate used to express that CFU is active at a subscriber. Such a predicate is necessary because we have left out existential quantifiers from the logic. The new predicate cfu_on satisfies:

$$CFU_{y \mapsto z} \Rightarrow cfu\_on(y) \qquad (26)$$

Call Forwarding is intuitively an extension to BC, i.e. already at the conception of CFU one thinks in terms of new rules. The rules are fairly obvious:

$$CFU_{y \mapsto z} : ready(x) \wedge idle(z) \xrightarrow{dial(x,y)} calling(x,z) \quad \text{if} \neq (x,y,z) \qquad (27)$$

$$CFU_{y \mapsto z} : ready(x) \wedge busy(z) \xrightarrow{dial(x,y)} rejecting(x) \quad \text{if} \neq (x,y,z) \qquad (28)$$

It is easily seen that the proposed specification is not operate-alone realizable, the new axioms are in conflict with the BC axioms. For instance it can be demonstrated that if there are at least three subscribers, A, B, and C, then any model of BC must have a state satisfying

$$ready(A) \wedge busy(B) \wedge idle(C)$$

If a $dial(A, B)$ event is executed in such a state, both the BC rule 13 and the CFU rule 27 specify the effect. The final state must consequently satisfy

$$rejecting(A) \wedge calling(A, C)$$

But this is impossible according to the BC network properties 6, 7 and 4.

In order to get rid of such conflicts we propose to weaken the relevant BC rules by ignoring those states where a new CFU rule is triggered. In particular we constrain rule 12 respectively rule 13 into:

$$\neg cfu\_on(y) : ready(x) \wedge idle(y) \xrightarrow{dial(x,y)} calling(x,y) \qquad (29)$$

$$\neg cfu\_on(y) : ready(x) \wedge busy(y) \xrightarrow{dial(x,y)} rejecting(x) \qquad (30)$$

To sum up, CFU consists of the following axioms:

BC axioms, old and modified	new CFU axioms
4–11, 14–22, 29, 30	26, 27, 28

As usual we tacitly assume that this CFU specification is operate-alone realizable.

7.3 Call Forward on Busy

The CFB feature allows a subscriber to forward incoming calls when he is busy. We use $CFB_{y\mapsto z}$ to formalize that y wants calls to be forwarded to z when he is busy.

The CFB is quite analogous to CFU. Consequently we will not go through a detailed development of its specification, we just give the resulting specification.

As for CFU we introduce a predicate which holds if a subscriber has activated CFB:

$$CFB_{y\mapsto z} \Rightarrow cfb\_on(y) \tag{31}$$

The rules for CFB then are:

$$CFB_{y\mapsto z} : ready(x) \wedge busy(y) \wedge idle(z) \xrightarrow{dial(x,y)} calling(x,z) \quad \text{if} \; \neq (x,y,z) \tag{32}$$

$$CFB_{y\mapsto z} : ready(x) \wedge busy(y) \wedge busy(z) \xrightarrow{dial(x,y)} rejecting(x) \quad \text{if} \; \neq (x,y,z) \tag{33}$$

Again we have to constrain those BC rules which specify the effects of the $dial$ event in the same cases. Rule 13 respectively rule 22 are consequently substituted with:

$$\neg cfb\_on(y) : ready(x) \wedge busy(y) \xrightarrow{dial(x,y)} rejecting(x) \quad \text{if} \; x \neq y \tag{34}$$

$$idle(x) : \neg(CFB_{z\mapsto x} \wedge busy(z)) \xrightarrow{dial(y,x)} true \quad \text{if} \; \neq (x,y,z) \tag{35}$$

To sum up, CFB consists of the following axioms:

BC axioms, old and modified	new CFB axioms
4–12, 14–21, 34, 35	31–33

We will henceforth assume that this specification of CFB is operate-alone realizable.

7.4 Call Waiting

The Call Waiting feature allows a subscriber to get a notification if he receives a call when speaking with someone else. He can then accept the new call and put the former party on hold by doing a flash hook. He can continue to swap between the parties by doing further flash hooks. In the end, if the subscriber does an on hook while a party is still on hold, he will be rung back from the held party.

It is more difficult to specify Call Waiting than it was for the previously introduced features. One reason is that introduction of CW makes the BC concept $path(x,y)$ deficient. Another is that the CW feature controls more phases of a call than the previously described features do.

Like for BC we will identify a number of phases which a CW invocation passes through. The phases will once again be formalized by predicates parameterized by the involved parties.

The first CW phase, $cw\_alerting(x,y,z)$, is entered when z calls x while x is talking with y. If x accepts the call from z by doing a flash hook, the $cw\_hold(x,z,y)$ phase will be entered where x is talking with z while y is on hold. Further flash hooks from x will keep the parties in the cw_hold phase, only the roles of the parties will change. Finally a new phase, $cw\_rejecting(x,\cdot)$, is necessary to cope with the situation which arises if the active partner of x does an on hook while the other partner is still on hold.

It is necessary to clarify how these phases affect the externally observable state of the system. For this we introduce another set of new predicates.

First of all a CW subscriber should somehow be notified that a third party is calling. For this we use the predicate $cw\_ringing(x)$ which indicates that x can hear a notification tone.

A predicate $on\_hold(x, y)$ is needed to express that x has y on hold.

We also have to realize that $engaged(x)$—the BC state where the telephone at x is currently used for communication with another party—must be split into two new states: When a party has been put on hold his telephone is $mute$, otherwise it is $responsive$:

$$engaged(x) \Leftrightarrow responsive(x) \vee mute(x) \qquad (36)$$

The two new states are furthermore mutually exclusive:

$$\neg(responsive(x) \wedge mute(x)) \qquad (37)$$

Notice that this implies that condition 4 also holds if $engaged(x)$ is replaced with $responsive(x)$, $mute(x)$.

Similarly we have to refine the conception of the phase $path(x, y)$. It is now possible that there is a communication path between two parties, even though they cannot speak with one another:

$$on\_hold(x, y) \Rightarrow path(x, y) \qquad (38)$$

We will provide an explicit name for the opposite situation: If there is a path between x and y and they can talk with one another, then $connected(x, y)$ holds. Obviously $path$ and $connected$ are related by:

$$connected(x, y) \Rightarrow path(x, y) \qquad (39)$$

Like $path$, the new predicate is furthermore symmetric:

$$connected(x, y) \Leftrightarrow connected(y, x) \qquad (40)$$

Connected parties have responsive telephones:

$$connected(x, y) \Rightarrow responsive(x) \wedge responsive(y) \qquad (41)$$

A possibly surprising fact is that we cannot outright require that $cw\_hold(x, y, z)$ implies $connected(x, y)$. The problem is that if y also subscribes to CW, then y may have put x on hold.

(An aside: Our initial formalization of CW actually contained the implication $cw\_hold(x, y, z) \Rightarrow connected(x, y)$. We soon had to realise, however, that this implication is too strong since it prevents CW from becoming operate-freely realizable, i.e. it causes CW to interact with itself. We have chosen not to further discuss this example detection since it is questionable whether the unrealizability is caused by a real interaction and not just by a specification error.)

The problem is solved by realizing that a CW subscriber cannot force a connection through, he can only accept it. That x accepts connection with y is formalized by the predicate $open(x, y)$. Two parties are connected, just in case they both accept:

$$open(x, y) \wedge open(y, x) \Leftrightarrow connected(x, y) \qquad (42)$$

If one party accepts connection, but isn't connected, then his telephone is mute:

$$open(x, y) \wedge \neg connected(x, y) \Rightarrow mute(x) \qquad (43)$$

The different phases of CW can now be further explained by using the introduced predicates:

$$cw\_alerting(x,y,z) \Rightarrow open(x,y) \wedge cw\_ringing(x) \wedge ring\_back\_tone(z) \qquad (44)$$

$$cw\_hold(x,y,z) \Rightarrow open(x,y) \wedge on\_hold(x,z) \qquad (45)$$

$$cw\_rejecting(x,y) \Rightarrow busy\_tone(x) \wedge on\_hold(x,y) \qquad (46)$$

With the provided explanatory formulas for the phases of CW the intuitive conception of the feature can now be formalized by the following set of rules.

$$CW_x : path(x,y) \xrightarrow{dial(z,x)} cw\_alerting(x,y,z) \text{ if } \neq (x,y,z) \qquad (47)$$

$$CW_x : cw\_alerting(x,y,z) \xrightarrow{flash\_hook(x)} cw\_hold(x,z,y) \qquad (48)$$

$$CW_x : cw\_hold(x,y,z) \xrightarrow{flash\_hook(x)} cw\_hold(x,z,y) \qquad (49)$$

$$CW_x : cw\_hold(x,y,z) \xrightarrow{on\_hook(x)} calling(z,x) \wedge rejecting(y) \qquad (50)$$

$$CW_x : cw\_hold(x,y,z) \xrightarrow{on\_hook(z)} open(x,y) \wedge idle(z) \qquad (51)$$

$$CW_x : cw\_hold(x,y,z) \xrightarrow{on\_hook(y)} idle(y) \wedge cw\_rejecting(x,z) \qquad (52)$$

$$CW_x : cw\_rejecting(x,y) \xrightarrow{flash\_hook(x)} open(x,y) \qquad (53)$$

$$CW_x : cw\_rejecting(x,y) \xrightarrow{on\_hook(x)} calling(y,x) \qquad (54)$$

Notice how some of the BC phases are reused in this specification. Furthermore notice that once again some of the effects of CW are deliberately left open, e.g. the effect of y doing an on hook in the phase $cw\_rejecting(x,y)$.

We must also indicate that the new phases should not be affected by certain actions. We propose the following rules:

$$cw\_alerting(x,y,z), cw\_hold(x,y,z) : true \xrightarrow{\substack{off\_hook(u), dial(u,v), \\ on\_hook(u), flash\_hook(u)}} true \qquad (55)$$
$$\text{if } u \notin \{x,y,z\}$$

$$cw\_rejecting(x,y) : true \xrightarrow{\substack{off\_hook(u), dial(u,v), \\ on\_hook(u), flash\_hook(u)}} true \text{ if } u \notin \{x,y\} \qquad (56)$$

Similarly, since we have introduced the new action *flash_hook*, we also have to specify that this action cannot affect the old BC phases which only involve other parties. This amounts to three new rules corresponding to rules 19–21 but with only *flash_hook* as trigger event where the argument of *flash_hook* should be the same as the argument of *off_hook* in rules 19–21. For brevity we omit the rules and just refer to them using primes: 19', 20', and 21'.

As it was the case for the previous features, the introduction of new CW axioms once again forces us to weaken BC.

Rule 13 describes how BC behaves when someone is calling an engaged party. CW is specifically designed to cope with this case, so—by also using principle †—rule 13 must be weakened to exclude just this case:

$$ready(x) \wedge busy(y) \wedge \neg(CW_y \wedge engaged(y)) \xrightarrow{dial(x,y)} rejecting(x) \qquad (57)$$

Similarly CW treats the on hook event in a way that differs from the BC rule 17. In order to weaken the rule we first have to express that a telephone is involved in a CW invocation. This is done with the new predicate cw_party:

$$cw\_alerting(x,y,z) \lor cw\_hold(x,y,z) \Rightarrow \\ cw\_party(x) \land cw\_party(y) \land cw\_party(z) \qquad (58)$$

$$cw\_rejecting(x,y) \Rightarrow cw\_party(x) \land cw\_party(y) \qquad (59)$$

Then rule 17 may be substituted with

$$path(x,y) \land \neg cw\_party(x) \land \neg cw\_party(y) \xrightarrow{on\_hook(x)} idle(x) \land rejecting(y) \qquad (60)$$

There is a final question to raise: Since $path$ has been refined using $connected$, what is the real intention behind the original BC axioms that use $path$? The axioms in question are the properties 8, 9 and the rules 15, 18, 20 20', 60. A bit of reflection shows that only the rules may need revision. It is obvious that $path$ must be substituted with $connected$ in rule 15 to ensure that the two parties really get connected. The resulting rule is denoted 15'. Rule 18 goes unchanged. We will require that rule 20 should express the same kind of frame axiom for $open(x,y)$ as it does for $path(x,y)$. The rule resulting from this extension is denoted 20''. Rule 20' also goes unchanged. It is less obvious what to do with rule 60; we propose to leave it as it is.

To sum up, CW consists of the following axioms:

BC axioms, old and modified	new CW axioms
4–12, 14, 15', 16, 18–19, 20', 21–22, 57, 60	19', 20'', 21', 36–56, 58, 59

8 Example detections

8.1 Interaction between $CFU_{B \mapsto C}$ and $TCS_{C:A}$

One interaction between CFU and TCS is easily detected. Assume that B is forwarding all incoming calls to C and that C is screening calls from A, i.e. each state of any model must have activation record:

$$AR = \{CFU_{B \mapsto C}, TCS_{C:A}\}$$

Furthermore all phones are assumed to be idle. This implies that any model must have an initial state that satisfies:

$$CFU_{B \mapsto C} \land TCS_{C:A} \land idle(A) \land idle(C)$$

Now A calls B. This means that A first does an $off\_hook(A)$ and next a $dial(A,B)$. From the axioms 11, 21 and the fact that all states of the model are assumed to have activation record AR we deduce that the first of these two events must lead to a state satisfying:

$$CFU_{B \mapsto C} \land TCS_{C:A} \land ready(A) \land idle(C)$$

According to axiom 27 the $dial(A,B)$ event then leads to a state satisfying:

$$TCS_{C:A} \land calling(A,C)$$

But according to axiom 23 this state must also satisfy:

$$\neg calling(A,C)$$

Since these two requirements are in direct conflict, no such model can exist. Consequently CFU and TCS are not cooperate-alone realizable—they interact.

8.2 Interaction between $CFU_{B \mapsto C}$ and $TCS_{B:A}$

Even though the involved features are the same as in the previous example the detected interaction is very different.

Assume the activation record $AR = \{CFU_{B \mapsto C}, TCS_{B:A}\}$. From an initial state with idle phones we can once again reach a state where A has become ready. According to axiom 25 a subsequent $dial(A, B)$ event will now lead to a state satisfying

$$rejecting(A)$$

But according to axiom 27 this state must also satisfy:

$$calling(A, C)$$

Now axioms 6 and 7 give that the state must satisfy:

$$ring\_back\_tone(A) \land busy\_tone(A)$$

But axiom 4 says that this is impossible. This shows another reason why CFU and TCS are not cooperate-alone realizable.

8.3 Interaction between $CFU_{B \mapsto C}$ and $CFB_{B \mapsto D}$

Next we treat the obvious interaction between two forwarding features where calls are forwarded to two different targets. Assume that B uses CFU to forward incoming calls unconditionally to C, but that he has also activated the CFB feature so that when B is busy, incoming calls are forwarded to D. This is captured by the activation record:

$$\{CFU_{B \mapsto C}, CFB_{B \mapsto D}\}$$

From the initial state of any model it is possible to arrive in a situation where B is busy, C and D are idle, and A is ready to dial:

$$ready(A) \land busy(B) \land idle(C) \land idle(D)$$

Now A dials the number of B. According to CFU, axiom 27, the resulting state satisfies $calling(A, C)$. But according to CFB, axiom 32, this state must also satisfy $calling(A, D)$.

Finally, assume that C does an off hook. This causes A and C to become connected, $path(A, C)$. But according to BC, axiom 20, this event also preserves $calling(A, D)$. Any model must consequently contain a state satisfying:

$$path(A, C) \land calling(A, D)$$

By using BC, axioms 8, 6, and 4, we see that these two requirements are in conflict, i.e. CFU and CFB interact.

Notice that the interaction is only detected in the step that *follows* A's call to B. The dial event in itself does not cause the conflict, in spite of the fact that it causes A to be calling C and D simultaneously.

8.4 Interaction between CW_B and $CFB_{B \mapsto C}$

Call Waiting and Call Forward on Busy are two features which are both designed to cope with a busy state at the subscriber. This causes an interaction when someone calls a subscriber of both features when this subscriber is busy.

Assume that B subscribes to CW, and also to CFB with calls forwarded to C:

$$AR = \{CW_B, CFB_{B \mapsto C}\}$$

By using BC axioms it is possible to show that any model for $(CW \cup CFB, AR)$ must have a state that satisfies:

$$ready(A) \wedge idle(C) \wedge path(B, D)$$

Now, assume that A dials B, and that C subsequently makes an off hook. We will show that the resulting situation is inconsistent.

According to the axioms 4, 8, 32, and 47, the first event causes the model to go into a state satisfying:

$$cw\_alerting(B, D, A) \wedge calling(A, C)$$

According to the axioms 55 and 15' the next event, C's off hook, then forces the model into a state satisfying:

$$cw\_alerting(B, D, A) \wedge connected(A, C)$$

According to the axioms 44, 39, and 8 such a state also satisfies:

$$ring\_back\_tone(A) \wedge engaged(A)$$

By 4 this is impossible and therefore CW and CFB are interacting.

Notice that as in the previous detection of interaction between CFU and CFB, the present interaction is likewise only detected in the step that follows the activation of the two features.

As already pointed out, CW and CFB are both designed to take care of the situation where the subscriber is busy. For this reason Bowen et al. call it a logical interaction [3]. This should not be taken to imply, however, that we cannot get the two features to cooperate. Actually the derived inconsistency suggests one way of resolving the interaction.

The interaction appears because some subscriber—in this case A—can't talk with one subscriber while calling another subscriber. However, this only indicates a limitation in the present telephone equipment. With a more advanced system, A's telephone could suppress the ring back tone while still calling the third party.

Such a resolution could even be desirable. Imagine e.g. a situation where a boss B subscribes to both CW and to CFB with calls forwarded to secretary C, and where B is talking to D. If A calls B now, B might still want to get an indication that A is calling and to have the opportunity of joining the ensuing conversation between A and C by doing a flash hook.

8.5 Interaction between CW_A and CW_B

In [4, Example 14 p. 334] an interaction between two instances of CW is presented. We now describe how our technique can be used to detect this interaction.

We assume that A and B have activated CW, and thus we have the activation record:
$$AR = \{CW_A, CW_B\}$$
It is possible to arrive in a situation where A and B have each other on hold. This could happen as follows: A first calls B; then C calls A and A does a flash hook to accept this call, thereby putting B on hold. While on hold, B then accepts a call from D and puts A on hold.

This preparatory scenario can be formally described using the rules of BC and CW. There are quite many details, however, so we only display the formula which the resulting state must satisfy:
$$cw\_hold(A, C, B) \wedge cw\_hold(B, D, A)$$

Now, assume that B makes an on hook event. From the axioms 50 and 51 it is seen that we must arrive in a state satisfying:
$$idle(B) \wedge calling(A, B)$$
But according to axioms 6 and 4 such a state also satisfies:
$$idle(B) \wedge \neg idle(B)$$

This is impossible. So (CW, AR) is unrealizable and we must conclude that CW is interacting with itself.

9 Future Research

We have described a method for interaction detection and demonstrated its applicability. There are a number of open issues, however.

The presentation has focused on semantical aspects; the described detections have mainly been guided by hand-waving. In order to automatize the detection process it is necessary to derive semantically sound deduction rules. Probably it will not be too difficult to find a convenient such set of rules.

The complementary problem of proving features free of interaction is probably more involved to tackle. For instance, in order to prove that two features are cooperate-alone realizable, we must find models for each activation record consisting of one active instance of each feature. This calls for a theoretical result showing that it is enough to consider a small (and finite!) number of typical situations. If such a result should come at hand, a *complete set* of the above deduction rules would make presence of interaction fully decidable. (Confront this with the undecidability result [8] where interaction is differently formalized.)

It would also be interesting to extend our underlying logic.

There is a need for introducing genuine variables and quantification. For instance quantification enables us to express the intention in property 26 more concisely:
$$(\exists z : CFU_{y \mapsto z}) \Leftrightarrow cfu\_on(y)$$

Actually the current development of CFU does not quite obey the proposed † principle as $cfu\_on(y)$ might be true without $CFU_{y \mapsto z}$ holding for any z. We have left the development as it is, however, because this possibility does not make the CFU specification more nor less realizable.

Another obvious possibility is to make the logic many-sorted. So far we have only presupposed one sort of constants. Thus there is no formal distinction between subscribers, telephones, directory numbers, and line numbers. In a more elaborate model of a network it would be natural to discern these concepts by letting each concept have its own sort. Then logical relations could express, for instance, that x is currently at telephone t.

Finally it could be convenient to express action refinement. The reader may have noticed that we have to specify the effect of a complete call by first specifying the effect of *off_hook* and then the effect of *dial*. Instead we propose to include the possibility of declaring one event to be a composition of other events, here:

$$call(x, y) = \mathit{off\_hook}(x); \mathit{dial}(x, y)$$

When using composite events in declarative transition rules, the intention is that if there is a path in a model labelled with the corresponding sequence of events, then the pre- and postconditions should be applied to the initial and final states in this path.

References

[1] Wiet Bouma and Han Zuidweg. Formal analysis of service/feature interaction using model checking. Technical Report TI-PU-93-868, PTT Research, Leidschendam, The Netherlands, February 1993. An abstract of this report may be found in: *International Workshop on Feature Interactions in Telecommunications Software Systems*, page 156. IEEE Communications Society, December 1992.

[2] Rezki Boumezbeur and Luigi Logrippo. Specifying telephone systems in LOTOS and the feature interaction problem. In *International Workshop on Feature Interactions in Telecommunications Software Systems*, pages 95–108. IEEE Communications Society, December 1992.

[3] T. F. Bowen, Ching-Hua Chow, F. S. Dworak, Nancy Griffeth, and Yow-Jian Lin. Views on the feature interaction problem. Technical Memorandum TM-ARH-012849, Bellcore, October 1988.

[4] E. Jane Cameron, Nancy Griffeth, Yow-Jian Lin, Margaret E. Nilson, Wiliam K. Schnure, and Hugo Velthuijsen. A feature interaction benchmark for IN and beyond. *Proceedings of FORTE '93*, pages 321–348. Also available as Technical memorandum TM-TSV-021982, Bellcore, 1992.

[5] E. Jane Cameron and Yow-Jian Lin. A real time transition model for analyzing behavioural compatibility of telecommunication services. In *Proceedings of the ACM SIGSOFT '91 Conference on Software for Critical Systems.*, pages 101–111. ACM Press, December 1991.

[6] Pierre Combes, Max Michel, and Beatrice Renard. Formal verification of telecommunication service interactions using SDL methods and tools. *Proceedings of the sixth SDL Forum*, pages 441–452, October 1993.

[7] M. S. Feather. Detecting interference when merging specification evolutions. In *Proceedings of the Fifth International Workshop on Software Specification and Design*, pages 169–176. ACM SIGSOFT, May 1989.

[8] Klaus Gaarder and Jan A. Audestad. Feature interaction policies and the undecidability of a general feature interaction problem. In *TINA '93*, pages II–189–II–200, 1993.

[9] Yasuaki Inoue, Kazumasa Takami, and Tadashi Ohta. Method for supporting detection and elimination of feature interaction in a telecommunication system. In *International Workshop on Feature Interactions in Telecommunications Software Systems*, pages 61–81. IEEE Communications Society, December 1992.

[10] Anthony Lee. Formal specification—a key to service interactions analysis. In *Proceedings of the Eigth International Conference on Software Engineering for Telecommunication Systems and Services*, 1992.

[11] John Mierop, Stefan Tax, and Ronald Janmaat. Service interaction in an object oriented environment. In *International Workshop on Feature Interactions in Telecommunications Software Systems*, pages 133–152. IEEE Communications Society, December 1992.

[12] Tadashi Ohta, Kazumasa Takami, and Akira Takura. Acquisition of service specifications in two stages and detection/resolution of feature interactions. In *Tina '93*, pages II: 173–187, September 1993.

[13] Pamela Zave. Feature interactions and formal specifications in telecommunications. *IEEE Computer*, 26(8):20–30, August 1993.

Using Temporal Logic for Modular Specification of Telephone Services [*]

Johan Blom, Bengt Jonsson, Lars Kempe [†]

Dept. of Computer Systems, Uppsala University, Box 325, S-751 05 Uppsala, SWEDEN

Abstract. We outline a methodology for the modular specification of telephone services within first-order linear-time temporal logic. Typically, the services offered by a telephone system consist of a basic service and several optional additional services, such as automatic callback, redirection, etc. We argue informally that temporal logic provides a flexible formalism for the specification of individual services, and for the composition of different services. We present a style of specification, in which the expected behavior of each additional service can be specified independently of other services. In this style, it is straight-forward to compose noninteracting services. We outline, by means of examples, how certain interactions between services that prescribe conflicting behavior can manifest themselves as inconsistencies when the services are composed. We then outline how the resolution of such interactions can be described in the formalism.

1 Introduction

The difficulty of designing and developing software for complex distributed computer systems, such as computer networks, telephone systems etc., is well-documented. It is a major obstacle in the development of e.g. services in telephone systems. Telecommunications software is huge, real-time, and distributed. It is difficult to add functionality to an existing system. Each new service (sometimes called a feature) may interact with many existing services, causing deviation from desired behavior or systems failure. The term *feature interaction* has been coined for this problem in the area of telephone systems [2].

One important problem is how to specify a telephone service in a structured way. The service offered by a telephone system typically consists of a basic service and several

---

[*]Supported in part by the Swedish Board for Industrial and Technical Development (NUTEK) as part of ESPRIT BRA project No. 6021 (REACT), and as part of grant No. 5321-93-3061 (on Feature Interaction).

[†]e-mail:{johan, bengt, larsk}@DoCS.UU.SE

optional additions such as automatic callback, redirection, etc. We will sometimes use the term *features* for these optional additions. The specification of such a telephone service should ideally be structured into modules, each module representing the basic service or an additional feature. The entire specification should be obtainable by composing these modules. However, the feature interaction problem makes it difficult to realize this aim of structuring a specification of a telephone system into understandable modules representing different features or services.

In this paper, we address the problem of structuring formal specifications of telephone services into modules that represent individual features. We aim for a specification style, in which each feature is specified in a separate module that specifies the expected behavior of the feature independently of other features. Ideally, a complete telephone service should be obtainable by composing a module containing a basic service with those modules that specify the desired set of features.

It seems that our aims are easiest to realize within a logically based specification language, such as Z [3, 10], where modules can be composed by logical operators. In the simple case, composition of individual features then corresponds to conjunction. During our work, we have found that our approach is less convenient to realize in languages such as LOTOS [1] and SDL [9], where the mechanism for composing parts of a specification is built on a certain mechanism for communication between modules. It is sometimes difficult to obtain the effect of simple conjunction in these languages.

In this paper, we have chosen first-order linear-time temporal logic (e.g. [7]) as a specification formalism. Our style of specification is inspired by the version of temporal logic presented in Lamport's TLA (Temporal Logic of Actions) [6]. Intuitively, TLA is based on the specification of individual operations or actions by a relation in first-order logic that describes the state change induced by this operation, like in Z. In addition, temporal logic provides operators for describing properties of entire executions. We argue that temporal logic can offer flexibility in expressing modules and the relationship between modules, due to the presence of logical operators for combining parts of a specification.

While our style of specification is inspired by TLA, our suggested style for specifying telephone services differs slightly from that of TLA. Intuitively, a feature is specified by describing how interactions with the environment (such as the dialling of a number) affect the abstract state of the telephone system. If features are independent, a specification of a complete service can then be obtained as the conjunction of the specifications that describe the desired features. For instance, assume that feature f_A prescribes an abstract state change specified by A in response to event e, and that feature f_B prescribes an abstract state change specified by B in response to event e. If A and B are consistent, the composition of these features will prescribe a change satisfying $A \wedge B$ in response to e. However, if f_A and f_B interact by prescribing conflicting responses to e, then this manifests itself as an inconsistency caused by the fact that $A \wedge B$ is equivalent to *false*. In this case, one or both of f_A or f_B must be transformed, e.g., by making them weaker. In the paper, we describe transformations such as this one in more detail.

Related Work As mentioned before, Lamport's work on Temporal Logic of Actions has been an inspiration for our work. Pnueli [8] describes how to use general linear-time temporal logic in a similar way as Lamport uses TLA.

Our specification style is presented using the part of temporal logic that specifies individual state-changes. This part is also present in Z [3, 10]. Thus, it would seem possible to present our ideas also in the context of Z, although we have not pursued this approach. An advantage of temporal logic is the possibility of expressing requirements as temporal properties of our specifications.

Related work that focuses on the feature interaction problem is by Inoue et al [5]. They use a rule-based framework for describing features, and describe a kind of interactions as the presence of conflicting rules that prescribe incompatible state changes.

The paper is organized as follows. In the next section, we present the basic semantic model used to model telephone services and also introduce temporal logic as the specification language for such models. The formal syntax and semantics of the temporal logic is not given in this paper. In Section 3 we outline a discipline of using temporal logic for specification of features in a telephone service. We first present a specification of a (simple) POTS in Section 4. Thereafter, in Section 5 the specifications of additional services are given, and in Section 6 we discuss the interactions between the services and POTS, and how to resolve them. Finally, in Section 7 we give conclusions and directions for future work.

2 Preliminaries

In this section, we introduce the formal model used to describe telephone services, introduce the use of temporal logic for specifying systems in this model, and present some notational conventions.

2.1 The Formal Model

As the basis for a formal specification of a telephone service, we use a formal model for the behavior of telephone systems. A telephone system is modeled as a reactive system. The state of the system is represented by a set of (abstract) *state variables* that describe the current state of the connections in the system. Interactions between the system and its users are represented by *external events*, e.g. the dialling of a number by a user.

A *state* of a system is an assignment of values to the state variables of a system. An event is the application of an *event predicate* to terms. We model the execution of a system as an infinite sequence σ of states of form

$$\sigma \triangleq s^0 \xrightarrow{events_1} s^1 \xrightarrow{event_2} s^2 \xrightarrow{event_3} s^3 \ldots$$

where the transitions between adjacent states may be labelled by external events. Intuitively, the external event labelling a transition represents the interaction that caused this transition. As an example, σ might be

$$\sigma \triangleq \{x \triangleq 0\} \xrightarrow{incr} \{x \triangleq 1\} \xrightarrow{incr} \{x \triangleq 2\} \xrightarrow{reset} \{x \triangleq 0\} \ldots$$

In this case, *incr* and *reset* are 0-ary event predicates. This system starts with the variable x being 0, and then changes x to 1 because of the external event *incr*. After this, x is again increased, this time to 2, also because of the external event *incr*. The external event *reset* then causes x to be set to 0.

A system is modeled by a set of infinite sequences, that represent the possible executions of the system.

2.2 Temporal Logic Specifications

Following the use of temporal logic presented by Lamport [6] and Pnueli [8], we specify the executions of a reactive system by temporal logic formulas that are built from the following parts.

Variables The state variables of the system are represented by variables in the specification. These variables are *flexible* in the sense that they can assume different values in different states.

Events Interactions between the system and its environment are represent by *events*. An event is an atomic formula, which is constructed as the application of a flexible *event predicate* to terms. Intuitively, an event is true in a state of the computation if the interaction represented by the event takes place in the following state transition. For example, *Dialling* can be an event predicate with two arguments. Intuitively, *Dialling*("Alan", 5432) is true in a state where user Alan dials number 5432 at the following transition between states[1].

Initial Condition An *initial condition*, which is an ordinary predicate in first-order logic, specifies the possible initial states of the system. An example is

$$Initial \triangleq x = 1 \land y = 2 \; ,$$

where *Initial* specifies that the variables x and y have 1 and 2 as initial values in any execution.

Reactive Part The possible state changes of the system are specified by *actions*. An action is a formula which relates the values of the state variables before the change to their values after the change. For this purpose, the operator $'$ is introduced. For any variable x, the value of x in the next state is denoted by x'. As an example, the action

$$x + 1 > x' \land y' = \neg y$$

states that in the next state, the variable x has to be less than the current value of $x + 1$, and that the variable y is negated in the next state. Note that this action does not state that variables other than x and y remain unchanged in the state transition, in contrast e.g. to the multiple assignment $x, y := x + 1, \neg y$ in some programming language. Actions can also specify the relation between the occurrence of events and state changes. As an example, the state changes in the example at the end of Section 2.1 satisfy the formula

$$incr \Rightarrow x' = x + 1$$

---

[1] Event predicates can be introduced without introducing flexible predicates into the logic, by assuming that events are abbreviations for actions (see below) that do not affect the remaining specifications.

The $\Box$ (always) operator is used to specify that the specification of a state change should apply to all state changes in an execution. If we again look at the example in Section 2.1 we conclude that something like

$$\Box \, [incr \Rightarrow x' = x + 1]$$

is true for σ since $incr$ occurs only when x is increased.

Sort Restriction For a tuple $x_1, \ldots, x_n$ of state variables, let $\text{Unchanged}(x_1, \ldots, x_n)$ be a shorthand for $\langle x_1, \ldots, x_n \rangle = \langle x'_1, \ldots, x'_n \rangle$, i.e., meaning that the variables $x_1, \ldots, x_n$ are not changed in the state transition. An action of form $\neg \text{Unchanged}(x_1, \ldots, x_n) \Rightarrow A$, where A is an action, then asserts that the variables $x_1, \ldots, x_m$ can only be changed if the action A is true. We call it a *sort restriction*, due to the fact that we will most often let A be the disjunction of a set of events; the action $\neg \text{Unchanged}(x_1, \ldots, x_n) \Rightarrow A$ then states that the variables $x_1, \ldots, x_n$ can only be changed by the events in A.

These building blocks may now be used to specify a reactive system by composing them with conjunction into a single formula that constitutes the whole specification. An example of this may be the following. If we introduce the abbreviations

$$\begin{aligned} Initial &\triangleq x = 0 \\ Reset &\triangleq reset \Rightarrow x' = 0 \\ Incr &\triangleq incr \implies x' = x + 1 \end{aligned}$$

then a system may be specified by the formula

$$Initial \wedge \Box \, [Reset \wedge Incr] \wedge \Box \, [\neg \text{Unchanged}(x) \Rightarrow (reset \vee incr)]$$

The event $reset$ allows a user to reset x to 0 at any time, and $incr$ allows a user to increment x.

2.3 Notational Conventions

The binding powers of the logical operators are (from strongest to weakest):

$$\neg, \; \wedge \;\; \vee \,, \Rightarrow \Leftrightarrow, \forall \, \exists$$

Parentheses and brackets can be used to change the precedence orders. If possible, we use standard mathematical notation in formulas. Some notation either originates from Z or is special for this paper, and is given below for clarity.

$S \triangleleft R$ denotes the pairs in relation R whose first elements are in set S.

$R \triangleright S$ denotes the pairs in relation R whose second elements are in set S.

$S \ntriangleleft R$ denotes the pairs in relation R whose first elements are not in set S.

$R \ntriangleright S$ denotes the pairs in relation R whose second elements are not in set S.

$R[s \mapsto t]$ denotes the relation obtained by deleting from R any tuple with s as the left element and adding the tuple $\langle s, t \rangle$. It can be defined as

$$(\{s\} \ntriangleleft R) \cup \{\langle s, t \rangle\} \;\; .$$

Any other notation used is explained in the context it occurs.

3 Specification Style

In this section, we describe the way in which we use temporal logic to write specifications of telephone services.

3.1 Specification Modules

Following the description in Section 2.2, a specification of a particular telephone service or feature consists of the following parts:

- A *set of state variables* that model the internal state of the service.

- A *set of event predicates* that model the interaction with the environment.

- An *initial condition* $ServiceInit$ which describes an initial internal state of the telephone service.

- An action $ServiceAct$ which specifies the restrictions on state transitions imposed by the service. In our specification style, $ServiceAct$ will (in most cases) be a conjunction of actions of the form

$$e \Rightarrow A$$

where e is an event and A is an action not containing events. This action intuitively describes the change of internal state caused by event e. We will sometimes use actions of form $e \Longleftrightarrow A$, in cases where the event e can be "identified" with the state changed specified by A.

- Possibly, a sort restriction $ServiceSort$.

A specification of the service can then be written as

$$Service \triangleq ServiceInit \land \Box \, [ServiceAct] \land \Box \, [ServiceSort]$$

3.2 Interactions between Features

Ideally, if a set of features are specified in the style just outlined, then their combination can be specified as the conjunction of specifications of the concerned features. However, when composing several features, it may happen that they interact by prescribing mutually inconsistent restrictions on the state changes in response to an event. The logical manifestation of such interactions is that the occurrence of the event is inconsistent with the specification, i.e., the specification may give rise to a deadlock if the environment is only willing to perform e. As a check for validating the specification, it is therefore advisable to check that interactions with the environment are enabled in the situations where we expect them to be.

For an action A, let $En(A)$ denote the formula

$$(\exists \overline{x}' : A)$$

where $\overline{x}'$ is a tuple of all flexible state variables in the specification. Intuitively, $En(A)$ is a property of states which expresses that it is possible to perform a state transition which satisfies the action A.

Suppose now that two features, $Service1$ and $Service2$ are defined by

$$Service1 \triangleq Service1Init \land \Box [Service1Act] \land \Box [Service1Sort]$$
$$Service2 \triangleq Service2Init \land \Box [Service2Act] \land \Box [Service2Sort]$$

If we want to check that these features are compatible with each other, in the sense that event e can always be performed, then we must check that a state change satisfying

$$e \land Service1Act \land Service2Act$$

can always be performed. In temporal logic, this can be expressed as checking that

$$Service1 \land Service2 \Rightarrow \Box [En(e \land Service1Act \land Service2Act)]$$

is true, i.e., that a telephone service satisfying the restrictions of both $Service1$ and $Service2$ may never arrive at a state where the occurrence of event e is inconsistent with the services. When more than two services are composed, or when we only require e to be enabled under certain conditions. the pattern is analogous (but more complex, of course).

In order to resolve the detected interaction, we must change the involved specifications, either or both of $Service1$ and $Service2$. We have investigated ways of changing either $Service1Act$ or $Service2Act$ to make the services compatible. One way of resolving the interaction is to identify the conditions on the state under which event e is unintentionally blocked, and decide that the restrictions of $Service2Act$ (say) should not apply then. This is like saying that $Service1$ has priority over $Service2$. If $cond$ specifies the conditions under which e is blocked, we can specify such a solution as

$$Service2 \triangleq Service2Init \land \Box [Service2Act \lor (e \land cond)] \land \Box [Service2Sort]$$

i.e., we relax the restrictions of $Service2$.

In the case that some totally new service, whose state changes are restricted by $NewAct$ (say), should be invoked as a result of the interaction, we can model the solution as

$$\begin{array}{l} Service1Init \land \Box [Service1Act \lor (e \land cond)] \land \Box [Service1Sort] \\ \land \quad Service2Init \land \Box [Service2Act \lor (e \land cond)] \land \Box [Service2Sort] \\ \land \quad \Box [(e \land cond) \Rightarrow NewAct] \end{array}$$

i.e., when e occurs in a state where $cond$ holds, then neither of the restrictions of $Service1$ or $Service2$ apply, and instead the restrictions of $NewAct$ apply.

4 The Basic Telephone Service

In this section, we illustrate the ideas outlined in Section 3, by applying them to the specification of the basic telephone service POTS (Plain Old Telephone Service), and a few additional features. We first outline how POTS can be described using the discipline. Section 5 describes the additional features Short Number, Hotline, Automatic

Recall and Call Forward on Busy. Some of them do not cause unintended interactions, whereas others cause interactions which must be resolved by some method.

We now present the specification of POTS according to the style presented in Sections 2.2 and 3. For the specification, we assume a set ID of users of the service, a set NO of telephone numbers, and a mapping $Users$ from NO to ID.

Variables   Define $POTSVars \triangleq \langle active, trying, busy, connected \rangle$, where

$active$: is a subset of ID, intuitively we have $id \in active$ if id has picked up his receiver.

$trying$: a subset of $ID \times ID$, intuitively we have $\langle id, id_2 \rangle \in trying$ if user id is trying to establish a connection with user id_2.

$busy$: a subset of $ID \times ID$, intuitively we have $\langle id, id_2 \rangle \in busy$ if user id has tried to establish a connection with user id_2 in a situation where id_2 has picked up his receiver (possibly but not necessarily connected to another subscriber)

$connected$: a set of unordered pairs of elements of ID. Intuitively we have $\langle id, id_2 \rangle \in connected$ if id and id_2 have established a connection in the system.

Events   Define $POTSEvents$ as

$$\exists id : Offhook(id) \vee \exists id, no : Dialling(id, no) \vee \exists id : Onhook(id)$$

where

$Offhook(id)$: denotes that user id lifts his receiver.

$Dialling(id, no)$: denotes that user id dials number no.

$Onhook(id)$: denotes that user id puts down his receiver.

Sort Restriction $POTSSort \triangleq \neg\text{Unchanged}(POTSVars) \Rightarrow POTSEvents$

Initial Condition $POTSInit \triangleq (active = trying = busy = connected = \emptyset)$

Reactive Part   A complete specification of the reactive part $POTSAct$ will not be given here, but only a part of it as an example. The complete specification is given in Appendix A. Here we only describe the restrictions imposed by $Dialling(id, no)$ in

POTS. First, define:

$$isActive(id) \triangleq (id \in active)$$

$$isBusy(id) \triangleq (isActive(id) \lor \exists id_2 : \langle id_2, id \rangle \in trying)$$

$$TryNotBusy(id, id_2) \triangleq \left[\begin{array}{l} \neg isBusy(id_2) \\ \land \quad trying' = trying[id \mapsto id_2] \\ \land \quad Unchanged(busy, connected) \end{array} \right]$$

$$TryBusy(id, id_2) \triangleq \left[\begin{array}{l} isBusy(id_2) \\ \land \quad busy' = busy[id \mapsto id_2] \\ \land \quad Unchanged(trying, connected) \end{array} \right]$$

$$TryCall(id, id_2) \triangleq \left[\begin{array}{l} TryNotBusy(id, id_2) \\ \lor \quad TryBusy(id, id_2) \end{array} \right]$$

Intuitively, $TryNotBusy(id, id_2)$ represents the addition of $\langle id, id_2 \rangle$ to $trying$, and $TryBusy(id, id_2)$ represents the addition of $\langle id, id_2 \rangle$ to $busy$, while preserving the values of other variables. We can now describe the effects of dialling a number by

$$\forall id, no : Dialling(id, no) \Rightarrow \left[\begin{array}{l} isActive(id) \\ \land \quad \neg \exists id_2 : \langle id, id_2 \rangle \in trying \\ \land \quad \neg \exists id_2 : \langle id, id_2 \rangle \in busy \\ \land \quad \neg \exists id_2 : \langle id, id_2 \rangle \in connected \\ \land \quad TryCall(id, FindBSubsc(no)) \\ \land \quad Unchanged(active) \end{array} \right]$$

This is a property of state changes, which states that if $Dialling(id, no)$ occurs at the state change, then id must have lifted his receiver, id must not be trying to call another subscriber already, and depending on the state of the subscriber that id calls, either the restrictions of $TryNotBusy(id, FindBSubsc(no))$ or of $TryBusy(id, FindBSubsc(no))$ are satisfied. Here, $FindBSubsc(no)$ is a function that returns the identity of the subscriber at no. It is originally defined in Appendix A. The function may be altered by other services, e.g. Call Forward in Section 5.4.

Specification of the Service

$$POTS \triangleq POTSInit \land \Box [POTSAct] \land \Box [POTSSort]$$

Validating the Specification To check that the specification does not unintentionally deadlock, we could check the formula

$$POTS \Rightarrow \Box \left[\begin{array}{l} \forall id, no : \left[\begin{array}{l} isActive(id) \\ \land \quad \neg \exists id_2 : \langle id, id_2 \rangle \in trying \\ \land \quad \neg \exists id_2 : \langle id, id_2 \rangle \in busy \\ \land \quad \neg \exists id_2 : \langle id, id_2 \rangle \in connected \end{array} \right] \\ \quad \Rightarrow En(POTSAct \land Dialling(id, no)) \end{array} \right]$$

for the event $Dialling(id, no)$ and similarly for other events.

5 Additional Services

This section describes the additional features Short Number, Hotline, Automatic Recall and Call Forward on Busy. The features are described informally and formally in the previously introduced style. The formal specifications may be skipped when first reading this section. Later, in the next section, the details of the services may be studied in parallel with the discussion about interactions.

5.1 Short Number Service

The short-number service makes it possible to perform the dialling event in two ways: either by dialling as normal, or by dialling a pre-programmed short number.

To the shortnumber service, we assume a set SNO of "short" telephone numbers, which in reality are separated from the normal (long) numbers, e.g. by a special control code such as #.

Variables

$shortdict$: is a subset of $ID \times SNO \times NO$ which acts as a partial function from $ID \times SNO$ to NO. Intuitively, the notation $shortdict(id, sno)$ stands for the (long) number represented by sno in the dictionary of user id, if one exists, and is defined to be $\bot$ if no such number exists.

Events are the following

$ReqShortNo(id, sno, no)$: occurs when subscriber id requests that shortnumber sno is to be interpreted as real (long) number no.

$CancelShortNo(id, sno)$: occurs when subscriber id requests that shortnumber sno is not a valid shortnumber anymore.

$DialLongNo(id, no)$: represents the usual way of dialling another subscriber.

$DialShortNo(id, sno)$: represents the dialling of short number sno by user id.

Sort Restriction

$$ShortNoSort \triangleq \neg \text{Unchanged}(shortdict) \Rightarrow \left[\begin{array}{l} \exists id, no, sno : ReqShortNo(id, no, sno) \\ \vee \exists id, sno : CancelShortNo(id, sno) \end{array} \right]$$

Initial Condition $\quad ShortNoInit \triangleq (\forall id, sno : shortdict(id, sno) = \bot)$

Reactive Part

$ShortNoAct \triangleq$
$\quad [\forall id, sno, no : ReqShortNo(id, sno, no) \Rightarrow shortdict' = shortdict[(id, sno) \mapsto no]]$

$\wedge [\quad \forall id, sno : CancelShortNo(id, sno) \Rightarrow shortdict' = shortdict[(id, sno) \mapsto \bot]\ ]$

$\wedge \left[\quad \forall id, no, sno : Dialling(id, no) \Leftrightarrow \left[\begin{array}{l} \left[\begin{array}{l} DialShortNo(id, sno) \\ \wedge no = shortdict(id, sno) \end{array} \right] \\ \vee\ DialLongNo(id, no) \end{array} \right] \right]$

The first two formulas in the definition of $ShortNoAct$ define the effect of the event predicates $ReqShortNo$ and $CancelShortNo$. The last formula defines the fact that $Dialling$ can be performed in one of two ways. Note that now some state transitions may be associated with more than one event, unlike in e.g. CSP [4]. We find this a natural consequence of the desire to refine events into different cases.

Specification of the Service

$\quad ShortNo \triangleq ShortNoInit \wedge \Box [ShortNoAct] \wedge \Box [ShortNoSort]$

5.2 Hotline Service

The hotline service works for special "hotline" subscribers. These subscribers only have to pick up the receiver for a call to be made to a predefined number. Typical hotline subscribers are emergency phones. The service handles the dialling to the hotline number just as POTS, so any redirection etc. applies as for normal dialling.

Variables There is only one variable, *hotlinedict*, which is a subset of $ID \times NO$ used to record the number that a certain subscriber is hotlined to. This variable implicitly records what subscribers have requested the service. If a subscriber is not hotlined to any number, then the service is not requested. We use the notation $hotlinedict(id)$ to denote the number that id is hotlined to. If id is not hotlined at all we define $hotlinedict(id)$ to be $\bot$.

Events The events $ReqHotline(id, no)$ and $CancelHotline(id)$ change the variable *hotlinedict* in analogy with the way that the short number service is requested and cancelled by the events $ReqShortNo(id, sno, no)$ and $CancelShortNo(id, sno)$. The **Initial Condition** (called $HotlineInit$), **Sort Restriction** (called $HotlineSort$) and **Specification of the Service** are also analogous to those of the short number service in Section 5.1.

Reactive Part $HotlineAct$ consists of restrictions concerning requesting and cancelling the service, together with the restriction

$\left[\forall id : \left[\begin{array}{l} OffHook(id) \\ \wedge\ isHotline(id) \\ \wedge\ \neg \exists id_2 : \langle id_2, id \rangle \in trying \end{array} \right] \Rightarrow TryCall(id, FindBSubsc(hotlinedict(id))) \right]$

where $isHotline(id)$ is defined as $\exists no : hotlinedict(id) = no$. This implication says that if an *OffHook* event occurs, the subscriber is a hotline subscriber, and no one is trying to call him, then we should try to connect to the hotline number.

5.3 Automatic Recall

This is the specification for the service that returns the last, not established, incoming connection when the receiver is picked up the first time. The following example shows how the service is used when C subscribes to the service:

- C calls A, but A is busy talking to B so C requests the Automatic Recall service.
- When A perform *OnHook*, C's telephone will start to ring for some amount of time, decided by the action *RingSignal*.
- When C picks up the receiver again, that is the external event *OffHook* occurs, the service will automatically try to establish a connection from C to A.

Variables

recallon: is used to record invocations of a request of the service.

recalldict: is a subset of $ID \times NO$ used to record the id of a subscriber that will automatically be called upon a new *OffHook* event. We use the notation $recalldict(id)$ to denote that id is automatically recalled to number no. If id is not Automatic Recalled at all we define $recalldict(id)$ to be $\bot$.

Events The events $ReqAutomaticRecall(id)$ and $CancelAutomaticRecall(id)$ change the variables *recallon* in analogy with the way that the short number service is requested and cancelled by the events $ReqShortNo(id, sno, no)$ and $CancelShortNo(id, sno)$. The **Initial Condition**, **Sort Restriction** and **Specification of the Service** are also analogous to those of the short number service in Section 5.1.

Reactive part The following definitions are used to define how the service behaves for different events from the user:

$$isAutomaticRecall(id) \triangleq \begin{bmatrix} id \in recallon \\ \wedge \ \exists no : recalldict(id) = no \end{bmatrix}$$

$AutomaticRecallAct \triangleq$

/* Restrictions concerning requesting and cancelling the service */

$$\wedge \begin{bmatrix} \forall id, no_1 : \begin{bmatrix} OffHook(id) \\ \wedge isAutomaticRecall(id) \\ \wedge no_1 = recalldict(id) \\ \wedge \neg \exists id_2 : \langle id_2, id \rangle \in trying \end{bmatrix} \Rightarrow \begin{bmatrix} recalldict' = \{id\} \triangleleft recalldict \\ \wedge TryCall(id, FindBSubsc(no_1)) \end{bmatrix} \end{bmatrix}$$

$$\wedge \begin{bmatrix} \forall id, no : \begin{bmatrix} OnHook(id) \\ \wedge isAutomaticRecall(id) \\ \wedge no = recalldict(id) \\ \wedge \neg isBusy(FindBSubsc(no)) \end{bmatrix} \Rightarrow RingSignal(id) \end{bmatrix}$$

$$\wedge \begin{bmatrix} \forall id, id_2, no : \begin{bmatrix} OnHook(id_2) \\ \wedge no = recalldict(id) \\ \wedge id_2 = FindBSubsc(no) \\ \wedge isAutomaticRecall(id) \\ \wedge \neg isBusy(id) \end{bmatrix} \Rightarrow RingSignal(id) \end{bmatrix}$$

$$\wedge \begin{bmatrix} \forall id, no : \begin{bmatrix} Dialling(id, no) \\ \wedge id_1 = FindBSubsc(no) \\ \wedge id \in recallon \\ \wedge isBusy(id_1) \end{bmatrix} \Rightarrow recalldict' = recalldict[id \mapsto no] \end{bmatrix}$$

Here, the first implication after the omitted restrictions on requesting and cancelling the service describes how the service is activated by trying to call a busy subscriber. The following two implications describe that user id is notified by the action $RingSignal(id)$ (which we do not specify here) that he may try again, and the last implication describes the actual recall.

5.4 Call Forward on Busy

This is the specification for the service that redirects an incoming trial to establish a connection if the subscriber is busy talking to someone else. It is possible to build chains of redirections.

Variables

$redirected$: is a subset of $ID \times NO$. Intuitively we have $\langle id, no \rangle \in redirected$ if subscriber id is redirected to the telephone number no.

Events The events $ReqCallForwardBusy(id, no)$ and $CancelCallForwardBusy(id)$ change the variable *redirected* in analogy with the way that the short number service is requested and cancelled, just as for the other features. The **Initial Condition, Sort Restriction** and **Specification of the Service** are also analogous to those of the short number service in Section 5.1.

Reactive part We give the full reactive part of the Call Forward on Busy service, even though it only consists of requesting and cancelling the service.

$$isCallForward(no) \triangleq \exists no : redirected(id) = no$$

$CallForwardBusyAct \triangleq$
$[\quad \forall id, no : ReqCallForwardBusy(id, no) \Rightarrow redirected' = redirected[id \mapsto no]\quad]$

$\wedge [\quad \forall id : CancelCallForwardBusy(id) \Rightarrow redirected' = \{id\} \triangleleft redirected \quad]$

These restrictions are orthogonal to $POTS$, since they only specify changes to the new variable *redirected*, and the service causes no interaction with $POTS$. This is mainly because of the chosen structure of $POTS$. See the section on interactions, section 6.2 for more details.

6 The Complete Telephone System

Our main philosophy is that additional features should be described simply by formulating their restrictions on possible state changes, simply as for POTS. The additional services may refer to the predicates, variables, and also events that were defined for POTS.

In this section, we describe how a complete telephone system is composed from the specifications described earlier, and also how we can detect interactions and resolve them.

We start by composing all services into a telephone system with all additional features. The system is defined as:

$$
\begin{array}{lllll}
& POTSInit & \wedge\ \Box[POTSAct] & \wedge\ \Box[POTSSort] \\
\wedge & ShortNoInit & \wedge\ \Box[ShortNoAct] & \wedge\ \Box[ShortNoSort] \\
\wedge & HotlineInit & \wedge\ \Box[HotlineAct] & \wedge\ \Box[HotlineSort] \\
\wedge & AutomaticRecallInit & \wedge\ \Box[AutomaticRecallAct] & \wedge\ \Box[AutomaticRecallSort] \\
\wedge & CallForwardInit & \wedge\ \Box[CallForwardAct] & \wedge\ \Box[CallForwardSort]
\end{array}
$$

This specification of the system has some serious flaws. For example: it is not possible for a user to perform normal dialling in all desired situations since some of the services interfere.

6.1 Noninterfering Features

One natural form of additional service is one which may add new state variables and/or events, and add additional restrictions on the state variables of the underlying service. Typically, the additional service may

- Extend the state by additional state variables that do not affect the existing state variables. An example of this is a ring signal service. We model the ring signal by the state variable *ringing* which is a subset of ID. Assume that we want to have a ring signal when someone is calling. The restrictions on changes to the signal are:

$$RingAct \triangleq (\forall id : id \in ringing \Leftrightarrow \neg isActive(id) \wedge \exists id_2 : \langle id_2, id \rangle \in trying)$$

It is obvious that we can define the addition of ring signals to the specification by conjoining $\Box$ $[RingAct]$ to a specification.

- Give a new interpretation to some event by introducing an equivalence between the event and some conditions. This strengthens the conditions for which an event may occur, and the requirement on the strengthening is that undesired deadlocks are avoided. An example of such a service is the *ShortNo* service. The service just states that we give *Dialling* another interpretation, such that when we dial a short number, this is interpreted as a normal dialling, but the short number has been translated into a long number.

6.2 The Additional Services and their Interactions

The remaining services Hotline, Automatic Recall and Call Forward on Busy interfere with POTS and each other. The resolving is explained below for each service.

The Hotline Service

For the case of the *Hotline* service, we have an interaction when *Hotline* is conjoined with $POTS$. This is because when $OffHook(id)$ and $isHotline(id)$ are true, the services $POTS$ and $Hotline$ require different restrictions on the state change. $POTS$ wants

$$IdleOrAnswer(id)$$

to be true (*IdleOrAnswer* is true when a subscriber lifts his receiver and either answers a call or there is no waiting call), and *Hotline* wants

$$TryCall(id, FindBSubsc(hotlinedict(id)))$$

to be true, and these two are mutually exclusive. This is solved by defining the complete system as

$$POTSInit \land HotlineInit$$

$$\land \quad \Box \left[POTSAct \lor \exists id : \left(\begin{array}{l} OffHook(id) \\ \land \quad isHotline(id) \\ \land \quad \neg \exists id_2 : \langle id_2, id \rangle \in trying \end{array} \right) \right]$$

$$\land \quad \Box [POTSSort] \land \Box [HotlineAct]$$

$$\land \quad \Box [HotlineSort]$$

Here we have changed the part $\Box [POTSAct]$ by adding a disjunct, which is an interaction handler that ensures that *Hotline* will restrict the system correctly when the *Hotline* service has been requested by a subscriber, and he lifts his receiver.

The Automatic Recall Service

This service interacts with *POTS* for the *OffHook* event. This is solved by including a handler in the composition, analogously as for *OffHook*. There are also interactions with the *Hotline* service. These can be solved by a handler that gives the *AutomaticRecall* service a higher priority.

$$POTSInit \land AutomaticRecallInit \land HotlineInit$$

$$\land \quad \Box \left[POTSAct \lor \exists id : \left(\begin{array}{l} OffHook(id) \\ \land \quad \left(\begin{array}{l} isAutomaticRecall(id) \\ \lor \quad isHotline(id) \end{array} \right) \\ \land \quad \neg \exists id_2 : \langle id_2, id \rangle \in trying \end{array} \right) \right]$$

$$\land \quad \Box \left[HotlineAct \lor \exists id : \left(\begin{array}{l} OffHook(id) \\ \land \quad isAutomaticRecall(id) \\ \land \quad \neg \exists id_2 : \langle id_2, id \rangle \in trying \end{array} \right) \right]$$

$$\land \quad \Box [AutomaticRecallAct]$$

$$\land \quad \Box [POTSSort] \land \Box [AutomaticRecallSort] \land \Box [HotlineSort]$$

Note that the handler added to *POTSAct* states that both Automatic Recall and Hotline have priority over POTS when an *OffHook* occurs. Similarly, the handler added to *HotlineAct* gives Automatic Recall priority over Hotline in cases when both services can be activated.

The Call Forward on Busy Service

This service interacts with *POTS* because the function $FindBSubsc()$ does not allow redirection as defined originally. However, this can easily be solved with an extension of the existing $FindBSubsc$ in the following way:

$\exists no_2 : redirected(User(no)) = no_2 \quad \Rightarrow \quad FindBSubsc(no) = FindBSubsc(redirected(User(no)))$
$\neg\exists no_2 : redirected(User(no)) \quad\quad\quad \Rightarrow \quad FindBSubsc(no) = User(no)$

The function is redefined as a recursive function that can find the redirections of subscribers.

7 Conclusion and Future Work

In the paper, we have outlined aspects of a methodology for using temporal logic for the modular specification of telephone services. We have tried to illustrate that temporal logic is a flexible tool for modularization, since it is not built around a particular model of communication. Of course, the absence of agreed upon default mechanisms for communication and variable hiding implies that a price must be paid, such as explicitly stating which variables may be changed by which events or actions. These additions are to a large extent only a syntactic burden and do not add significantly to the intrinsic complexity of the specification. The absence of mechanisms for communication could become problematic when describing more concretely the architecture of a specific system design. In this paper we have specified only a high-level view of services.

The paper is part of an experiment in assessing the usefulness of temporal logic for structuring specifications. We believe that temporal logic with its flexibility is a suitable vehicle in such an experiment. We also hope that the insights gained can be transferred also to other frameworks.

8 Acknowledgments

We are grateful to Jan-Olof Nordenstam and his collaborators for valuable interaction (sic!) and discussions, and to Yih-Kuen Tsay for many insightful comments and discussions.

References

[1] T. Bolognesi and E. Brinksma. Introduction to the ISO specification language LOTOS. *Computer Networks*, 14(1), Jan. 1989.

[2] T.F Bowen, F.S Dworack, C.H. Chow, N. Griffeth, G.E. Herman, and Y-J Lin. The feature interaction problem in telecommunications system. *SETS*, 1989.

[3] I.J. Hayes. *Specification Case Studies*. Prentice-Hall, 1987.

[4] C. A. R. Hoare. *Communicating Sequential Processes*. Prentice-Hall, 1985.

[5] Yasuaki Inoue, Kazumasa Takami, and Tadashi Ohta. Method for supporting detection and elimination of feature interaction in a telecommunication system. In *International Workshop on Feature Interactions in Telecommunications Software Systems*, 1992.

[6] L. Lamport. The temporal logic of actions. Technical report, DEC/SRC, 1991.

[7] Z. Manna and A. Pnueli. *The Temporal Logic of Reactive and Concurrent Systems*. Springer Verlag, 1992.

[8] A. Pnueli. System specification and refinement in temporal logic. In R.K. Shyamasundar, editor, *Foundations of Software Technology and Theoretical Computer Science*, volume 652 of *Lecture Notes in Computer Science*, pages 1–38. Springer Verlag, 1992.

[9] R. Saracco, J.R.W. Smith, and R. Reed. *Telecommunication Systems Engineering using SDL*. North-Holland, 1989.

[10] J.M. Spivey. *The Z Notation*. Prentice-Hall, 1989.

Appendix Definition of the Reactive Part for POTS

We start by giving all definitions needed to specify the restrictions required by POTS on the telephone system.

$$isActive(id) \triangleq (id \in active)$$

$$isBusy(id) \triangleq (isActive(id) \lor \exists id_2 : \langle id_2, id \rangle \in trying)$$

$$StartIdle(id) \triangleq \left[\begin{array}{l} \neg \exists id_2 : \langle id_2, id \rangle \in trying \\ \land \quad \text{Unchanged}(trying, busy, connected) \end{array} \right]$$

$$Answer(id, id_2) \triangleq \left[\begin{array}{l} \langle id_2, id \rangle \in trying \\ \land \quad trying' = trying \setminus \{\langle id_2, id \rangle\} \\ \land \quad connected' = connected \cup \{\langle id, id_2 \rangle\} \\ \land \quad \text{Unchanged}(busy) \end{array} \right]$$

$$IdleOrAnswer(id) \triangleq \left[\begin{array}{l} StartIdle(id) \\ \lor \quad \exists id_2 : Answer(id, id_2) \end{array} \right]$$

$$TryNotBusy(id, id_2) \triangleq \left[\begin{array}{l} \neg isBusy(id_2) \\ \land \quad trying' = trying[id \mapsto id_2] \\ \land \quad \text{Unchanged}(busy, connected) \end{array} \right]$$

$$TryBusy(id, id_2) \triangleq \left[\begin{array}{l} isBusy(id_2) \\ \land \quad busy' = busy[id \mapsto id_2] \\ \land \quad \text{Unchanged}(trying, connected) \end{array} \right]$$

$$TryCall(id, id_2) \triangleq \left[\begin{array}{l} TryNotBusy(id, id_2) \\ \lor \quad TryBusy(id, id_2) \end{array} \right]$$

Then we define the restrictions on the system imposed by POTS. Essentially, we define one big predicate *POTSAct* that describes what state changes must occur for different conditions.

$POTSAct \triangleq$

$$\left[\quad \forall id : \mathit{OffHook}(id) \Leftrightarrow \begin{bmatrix} \neg isActive(id) \\ \wedge\, active' = active \cup \{id\} \end{bmatrix} \quad \right]$$

$$\wedge \left[\quad \forall id : \mathit{OnHook}(id) \Leftrightarrow \begin{bmatrix} isActive(id) \\ \wedge\, active' = active \setminus \{id\} \end{bmatrix} \quad \right]$$

$$\wedge [\quad \forall id : \mathit{Offhook}(id) \Rightarrow IdleOrAnswer(id) \quad]$$

$$\wedge \left[\quad \forall id, no : \mathit{Dialling}(id, no) \Rightarrow \begin{bmatrix} isActive(id) \\ \wedge \neg \exists id_2 : \langle id, id_2 \rangle \notin trying \\ \wedge \neg \exists id_2 : \langle id, id_2 \rangle \notin busy \\ \wedge\, TryCall(id, FindBSubsc(no)) \\ \wedge\, \text{Unchanged}(active) \end{bmatrix} \quad \right]$$

$$\wedge \left[\quad \forall id : \mathit{Onhook}(id) \Rightarrow \begin{bmatrix} trying' = \{id\} \triangleleft trying \\ \wedge\, busy' = \{id\} \triangleleft busy \\ \wedge\, connected' = (\{id\} \triangleleft connected) \triangleright \{id\} \end{bmatrix} \quad \right]$$

The function $FindBSubsc$ is used to find the id of the B-subscriber. It is originally defined as a function:

$$FindBSubsc(no) = User(no) \quad .$$

This is the original version of the function, and this function is changed by the addition of other services.

The Negotiating Agents Approach to Runtime Feature Interaction Resolution

Nancy D. GRIFFETH
Bellcore, 445 South Street, Morristown, NJ 07962-1910, USA

Hugo VELTHUIJSEN
PTT Research, St. Paulusstraat 4, 2264 XZ Leidschendam, The Netherlands

Abstract. This article describes how to use the Negotiating Agents approach on a telecommunications platform. Negotiation is used in this approach to resolve conflicts between features of different users and between different features of one user. The theory behind the approach is discussed briefly. Methods for implementing the approach are given along with the methods for defining IN features that allow negotiation to resolve conflicts between them.

1 Introduction

Rapid change in the telecommunications industry increases the complexity not only of building but also of using telecommunications services. Much of the complexity arises from the feature interaction problem. When features interact, a user must understand the behavior of features in combination – even how features of other users may affect the behavior of her features. Similarly, a service provider must determine how combinations of features will behave, including combinations of its own features with other providers' features. The need to understand features in combination, rather than individually, limits the ability of users to use the network and the ability of providers to add to it. In this paper, we propose an approach to one class of features and feature interactions that avoids the need to understand features in combination. This approach also provides an elegant mechanism for detecting and resolving feature interactions.

We assume that there is a standard platform underlying all services on a telecommunications network and offering a collection of operations to the users of the network. Even though this is not the case with existing networks, various groups are working to standardize such a platform [1, 6]. Given such a platform, we define two kinds of features. One kind of feature is a *technology* feature, which is an individual operation that the platform provides. A new technology feature is created by creating a new operation on the platform. We imagine that these features will usually correspond to new technology or new equipment provided in the network – for example, voice activated dialing or voice recognition technology. The second kind of feature is a *policy* feature, which is a constraint on the set of operations that a user or provider is willing to perform in initiating or modifying a call. For example, an end-user may say that his customer equipment should never be connected to a particular number, or a service provider may say that its service should be provided only to an enumerated list of subscribers.

The problem of interactions between technology features is quite different from the problem of interactions between policy features. We address the problem of interactions between policy features only, and henceforth in this paper the term "feature" refers to policy feature. A *single-party feature interaction* occurs when it is not possible to satisfy all of the constraints of a single subscriber. A *multi-party feature interaction* occurs when it is not possible to satisfy all of the constraints of all parties to an attempted call.

Today's features modify the process of call setup. The correct functioning of any such feature depends on assumptions about how this process works. But since what all features do is to modify this process, interaction between features is inevitable. The interaction is sometimes desirable (as when Block Calling Number Delivery prevents Calling Number Delivery from providing the originator's number to the terminating party) and sometimes undesirable (as when Call Forwarding allows calls to be set up that Call Screening features would prevent). In any approach to the problem of feature interactions, it is necessary to describe the intended effect of a feature on a call in order to distinguish between desirable and undesirable interactions.

Our approach is to enable users and providers involved in a call to specify directly what are the constraints on the calls they will accept, instead of having to embody these constraints in features that are hard to understand and hard to use. We assume that constraints are expressed in terms of operations that a user or provider is willing or unwilling to perform. To resolve conflicts between the constraints of different users, we provide a mechanism for them to *negotiate* with each other to determine what set of operations will be used to initiate or modify a call. Such a negotiation process provides an automated method for detecting and resolving policy feature interactions at run time. This negotiation method guarantees that the policies or intentions of the various entities are respected, while calls are set up whenever possible. An important consequence of the use of negotiation is that the autonomy of different users and providers can be preserved.

We claim that negotiation provides the following additional advantages to end users and to service providers. The body of the paper substantiates these claims.

- Single-party feature interactions are automatically detected. Resolution of these interactions is up to the user or service provider.

- Multi-party feature interactions are automatically detected and resolved at run-time.

- Each user and subscriber can specify an individual policy involving calls.

- All calls that can be set up without violating someone's policy will be set up.

- Unanticipated feature interactions (even those due to faulty implementation or to system failures not involving the negotiation mechanism) will be detected if they result in a violation of someone's policy.

- No one need be aware of any policy features belonging to anyone else, and in particular there is no need to know about the combinations of policy features used in setting up a particular call.

In section 2 of this paper, we present examples of features and their interactions for which more conventional methods of automatic feature interaction detection are problematic. In section 3, we describe how to incorporate our paradigm for negotiation in an arbitrary extensible, object-oriented platform for telecommunications. Our negotiation mechanism has been presented in detail elsewhere [4, 5], but we review it in section 4. Then we describe how negotiation has been implemented on one platform, the Touring Machine^{TM} platform [1]. In section 6, we discuss how some generic AIN and IN CS-1 features would be designed using our paradigm. Finally, in section 7, we summarize the advantages and discuss some remaining research issues.

2 Example Feature Interactions

We present features in this section which subvert the intention of another user's features. This kind of interference appears to require that each user's service "know", in some sense, about all features that may be encountered. Furthermore, it appears that in order to resolve the interaction, we must favor one feature over another. Our work offers an alternative to knowing about other features and also to favoring one over another. We use two examples to illustrate the basic idea behind our approach.

One example of a feature which can subvert the intended purpose of another feature is calling number delivery, which provides the number of the originating subscriber to the terminating subscriber. This feature can interact with the unlisted number feature. Whether it does or not depends on the subscriber's reason for subscribing to unlisted number. If the subscriber doesn't want her number to be known, calling number delivery subverts her intentions. This example and our resolution of it is discussed in detail in [4, 5]. We assume that the subscriber to unlisted number is willing to identify herself, but not to provide her number, and that the subscriber to calling number delivery wants only to know the identity of the calling party. Our proposal to satisfy the intentions of both parties to the call is to deliver the name instead of the number.

In order to illustrate the application of negotiation more fully, we present a second example here which raises similar issues. Suppose that one user has the Terminating Key Code Protection (TKCP) feature, which protects the line by requiring any caller to enter a user-defined key before the call is offered to the line.

Consider the interaction between TKCP and Call Forwarding (CF). First, suppose that the subscriber with Call Forwarding (called F from now on) has forwarded all calls to the line (called L) of the subscriber (called K) with terminating key code protection. Any call to F will be forwarded to L and the caller will be asked to enter a key, which he may not know. Even if he does know the key, he cannot use it unless he has been told which line the call has been forwarded to.

Appropriate resolution of the interaction requires knowing the intentions of the subscribers when they activated the features. The services mentioned above can be used for a number of different purposes. We select just two of these purposes for each feature for the sake of this example.

One purpose of call forwarding is to make sure that calls to a subscriber's number (e.g., her home phone) still reach her when she is elsewhere. A second is to redirect calls appropriately, as to a secretary or voice-mail.

Similarly, one purpose of Terminating Key Code Protection is to make sure that only authorized users are put through to the line. In other words, TKCP protects the line from unauthorized callers. Another purpose is to permit the subscriber to select which callers he will talk to. In this case, TKCP protects the subscriber from random callers.

---

^{TM}Touring Machine is a trademark of Bellcore.

Considering these purposes, let's suppose that a friend is calling F and F has forwarded calls to L because that's where F is. We argue that the correct resolution of the interaction is to require the key if the purpose of the key is to protect the line. But if the purpose of the key is to protect the subscriber, then the key should not be required when a different subscriber is the object of the call.

On the other hand, if F has forwarded calls to L because F wants K to handle the calls, then the key should be required whether TKCP protects the subscriber or the line.

In most cases the key will still be required, but we see that in one case it is possible to satisfy the goals of both subscribers and complete the call without requiring the key.

This approach is based on the following observation. Typically, the particular implementation of a feature does not represent the ultimate intention of a user, but merely one way to achieve a user's intentions. There might exist alternative ways to achieve these intentions as well. Such alternatives provide room for negotiation. To explore these alternatives, it is necessary for a user agent — after receiving a request to set up a call that is unacceptable to that user — to recognize what intention might be behind the request, and to derive from that intention alternative (possibly acceptable) ways to achieve it. Even when a user is not informed explicitly about the intentions of other users, the user may be able to speculate about the intention, based on the information he or she does have, i.e., the received request.

We automate the process of recognizing intentions by building a hierarchy of possible goals on top of primitive system operations. If a subscriber is not willing to agree to a proposed set of operations to initiate or modify a call, an agent of the subscriber uses the hierarchy to infer which goal was intended by the caller, and then checks to see if there is an alternative way to achieve the same goal. If there is an alternative, then instead of rejecting the call outright the subscriber can offer the alternative as a counterproposal.

3 Implementing Negotiation as Part of a Telecommunications System

A negotiating system negotiates about what collection of operations to perform on a given platform. We describe three levels in a telecommunications system that uses negotiation: the platform itself, the negotiating objects, and the user interface.

3.1 The platform

Our mechanism for negotiating assumes that call initiation and modification are described by a collection of *operations* that determine the form of the call. Policy features and technology features are both related to these operations: the technology features determine which operations are available, and the policy features determine which operations a system user is willing to use. In order to add features rapidly and easily, including new operations, the operations should be provided by an extensible, object-oriented platform for telecommunications [1, 6]. We believe that the issues of interaction between technology features can be addressed by using such platforms.

Policy features determine which operations a system user is willing to execute, but given a collection of operations that have been proposed by one system user, how can

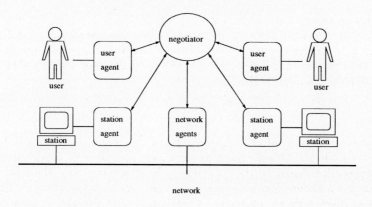

Figure 1: The logical structure of a negotiating system for telecommunications.

we decide which other system users or providers should be asked to authorize the operations? To do this, we require that for each instantiation of each operation's arguments, we can determine which system users or providers should authorize the operation. A user may want to authorize any operation that affects the user, for example, by connecting equipment she owns to other network resources or by billing her for use of network resources. Also, a provider may want to authorize any operations making use of the providers resources. Furthermore, we require that the collection of system users or providers that must authorize a collection of operations be the same as the union of the sets of authorizing users or providers for the individual operations.

A telecommunications system user creates a proposal for initiating or modifying a call as a partially ordered set of the operations that are provided by the platform. (The partial order refers to the order of execution of the operations.) The set of operations is sent to the negotiating system, which returns a possibly different set of operations to be executed. Corresponding to this set of operations is a set of system users and providers that must authorize it. The negotiation mechanism guarantees that the required users and providers have authorized any set of operations it returns.

3.2 The agents and negotiator

A negotiating system works on behalf of *entities*, which have policies restricting what collections of operations they are willing to perform. It includes *agent objects*, which represent the various entities in the system and try to carry out their policies, and *negotiators*, which help the agents reach agreement. Figure 1 illustrates the interrelationships in a negotiating system for telecommunications. In the figure, the entities include system users, local equipment (telephones, videophones, workstations, etc.), and the public network itself. Other potential entities would include information providers, software service providers, and various communications providers.

An agent object is assigned to each entity. The functions of an agent are to produce proposals to initiate or modify a call and to evaluate proposals from other agents, possibly generating counter-proposals. *Proposals* specify the desired operations on calls. If a proposal is not acceptable to an entity, the agent constructs a counter-proposal. In the following section, we describe a mechanism for evaluating and generating proposals and counter-proposals.

We distinguish between three quite different organisations for negotiation:

Direct negotiation. In this case, agents negotiate directly with each other without the assistance of a mediator.

Indirect negotiation. Here a dedicated entity is used to recognize which agents have to approve a proposal and to route proposals and counter-proposals to the appropriate agents. We call such an entity a *negotiator*. This entity could also be used to monitor the progress of a negotiation session, determine when agreement is reached, and even enter proposals into the negotiation process that have proven to be widely acceptable in previous sessions.

Arbitrated negotiation. An *arbitrator* takes the complete script of each agent and has sole responsibility for finding a resolution of a conflict. Thus, in arbitrated negotiation the agents don't need to generate and evaluate proposals.

We use a negotiator object for indirect negotiation between agents. There are several reasons to opt for this organisation. The advantages of indirect negotiation (and also arbitration) over direct negotiation are the following.

1. In contrast to direct negotiation, a separate negotiator can be used to monitor the negotiation process and to make sure that actual progress is being made (otherwise a non-terminating sequence of proposals and counter-proposals might be created).

2. The use of a mediator (negotiator or arbitrator) centralizes communication and makes it easier to include new agents and to communicate with those new agents.

3. The negotiator can use knowledge acquired by monitoring many previous negotiation sessions to propose solutions.

4. Different mediators can be used for different situations, providing benefits of specialization. For instance one can envision 'smart' and 'dumb' negotiators which provide more or less support and hence could be more or less expensive.

5. The mediator can act as an "honest broker," in that it prepares, based on proposals it has handled before, an agreement that is about equally good for all parties.

The primary advantage of either direct or indirect negotiation over arbitration is that it is more generally applicable. In many cases, including telecommunications systems, the parties to a negotiation may not want to share their preferences with any third party. It used to be that a phone company was seen as a trusted party, but with the current increase in competition between the phone companies makes them less likely candidates for this role.

Another reason to prefer a negotiator over an arbitrator is that a negotiator makes it possible for agents to distinguish themselves from other agents and to improve their results by being smarter than other parties. Actually, this may seen as both a good thing and a bad thing. In general, the problem of malevolent agents trying to take advantage of other system entities must be addressed.

It appears that arbitrated negotiation is computationally less expensive then indirect negotiation since all relevant information is locally available in the arbitrator.

However, the arbitrated-negotiation approach requires that *all* information is sent to the arbitrator, while in case of indirect negotiation information will be sent to some other entity only when it appears to be pertinent to resolving a particular conflict. This suggests that arbitrated negotiation is more expensive in terms of communication overhead. In future work, we may address the benefits and drawbacks of these different methods in more detail. For now, we concentrate on the case of indirect negotiation.

3.3 The user interface

System users and providers need access to two kinds of functionality: one for expressing constraints on the sets of operations they will agree to and the other for submitting proposals. The first functionality is like provisioning. This is how a user defines the policy part of her service. The second functionality initiates call set-up or modification.

A constraint language must be provided to describe which collections of operations can be allowed in which states. Such constraints can be expressed in any logic language, but to avoid intractability we have chosen a simple language based on a small collection of user goals. The user specifies the acceptability (or unacceptability) of each of these goals. In the next section, we describe how the acceptability of a proposal can be inferred from the acceptability of goals.

In this language, an entity must be able to specify a finite set of states that it can enter; enumerate the instantiated goals that are acceptable or unacceptable in each state; and specify the operations that trigger a transition from one state to another. For example, suppose that a platform offers two operations, connect(x,y) and hangup(x,y). A user might define two states, **busy** and **idle**, and say that connect(x,y) is acceptable in the idle state and not otherwise, while hangup(x,y) is acceptable in the busy state and not otherwise. The operation connect(x,y) triggers a transition from idle to busy state and hangup(x,y) triggers a transition from busy state to idle state.

When a user wants to initiate or modify a call, he simply needs to select a goal. The negotiating system will do what it can to find a way of meeting the goal.

4 Negotiation mechanism

We now present a description of our negotiation mechanism. The description of the negotiation mechanism uses the example of Terminating Key Call Protection and Call Forwarding from the preceding section. We use this example to show how the process of generating and evaluating proposals and counter-proposals can be formalized and automated. The process uses a *goal hierarchy*, a definition of what *acceptable* proposals are, and algorithms for determining acceptability of proposals and generating counter-proposals. For a more detailed description of the mechanism we refer the reader to [4, 5].

4.1 The goal hierarchy

The basis of negotiation is a *goal hierarchy* whose lowest-level (basic) goals are operations on a given platform. Higher-level goals correspond to possible goals or intentions of a user. Examples of basic goals in figure 2, containing the goal hierarchy for our example, are get-key(t,u) and connect(u,s,v,t). The operation get-key(t,u) asks

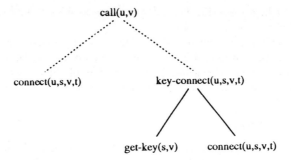

Figure 2: Goal hierarchy for our example negotiation process.

user u to enter the key for station t. It succeeds if u responds with the correct key. The operation connect(u,s,v,t) connects stations s and t on behalf of users u and v.

New goals can be formed by combining other goals in one of two ways. The goal call(u,v) is an *abstraction* of goals connect(u,s,v,t) and key-connect(u,s,v,t). The abstraction relation between connect(u,s,v,t) and call(u,v) with uninstantiated parameters is actually short hand and should be interpreted as follows: given an assignment of values, say U and V, to the parameters u and v, the goal call(U,V) is an abstraction of the goals connect-station(U,S,V,T) for any assignment of values S and T to parameters s and t.

Thus, returning to the goal hierarchy of figure 2, the goal call(u,v) is achieved if either connect(u,s,v,t) or key-connect(u,s,v,t) is achieved for some value of s and t. When one goal is an abstraction of a second goal, then we say that the second goal is a *specialization* of the first. Achieving a specialization of a goal implies achieving the goal itself.

We also define a *composition* relation between goals. In our example, the goal key-connect(u,s,v,t) is composed of get-key(t,u) and connect(u,s,v,t). A composite goal is achieved if all its component goals (or *subgoals*) are achieved. Based on these two types of relationships, we can build a hierarchy, as illustrated in figure 2. The nodes represent goals, the broken lines represent abstraction relationships, and the solid lines represent composition relationships.

Whenever a goal needs to be achieved, that goal can also be achieved by achieving *any* of the specializations of that goal or by achieving *all* of its composite subgoals. Thus, the abstraction and composition relations define the different ways that a goal can be achieved. We use these notions of abstraction and composition to define a specification of a goal. A *specification* of a goal is derived by recursively replacing an abstract goal by one of its specializations and a composite goal by the set of its components. Let home(x) stand for the location of the home (or usual) station for user x. In the example of figure 2, {connect(A,home(A),B,home(B))} is a specification for the goal call(A,B). This means that one way for A to call B, is to set up a connection between the home station for A and the home station for B. Another specification for the goal call(A,B) is {connect(A, S, B, T)}. This means that another way for A to call B is to connect stations S and T. This will work as intended only if A is at station S and B is at station T. We will discuss how to handle this in section 6. A third specification of the goal call(A,B) is {get-key(A,home(B)),connect(A,home(A),B,home(B))}. If this specification is used to set up the call, A will be required to produce the key protecting station home(B) before the connection will be made. By definition of a specification, whenever all goals in the specification of a goal are achieved, then that goal itself is achieved.

We use specifications as the proposals and counter-proposals in our negotiation mechanism. Whenever a proposal is received, the hierarchy can be used to infer what goals it is trying to achieve (for a detailed description of the algorithms see [5]). The following section defines what an acceptable proposal is.

4.2 Acceptability

A specification (and thus a proposal) is *acceptable* for an agent if the entity that the agent represents (e.g., a subscriber) would agree to it. A specification is *unacceptable* if the entity would not agree to it. Each specification is either acceptable or unacceptable to an agent. However, there are possibly many specifications for each goal, so a subscriber is required only to record the acceptability of goals (a subset of all specifications), not the acceptability of specifications. A goal can be marked either acceptable or unacceptable; or it can be left unmarked. When a goal is unmarked, we call that goal *indeterminate*. The acceptability of specifications can be derived from the acceptability of goals they achieve by using a number of rules, as is described below.

All specifications of an acceptable goal are acceptable. This means that by marking a goal as acceptable, a subscriber indicates that no matter how a goal is achieved, it will be acceptable. Analogously, no specification of an unacceptable goal is acceptable. A goal with no marking may have both acceptable and unacceptable specifications.

It can happen that a goal is unacceptable for an agent, while all its component goals are acceptable. For instance, a subscriber may agree to talk to either one of two other subscribers, but not to both at the same time. Likewise, a goal may be acceptable for an agent, while one of the component goals is unacceptable. For instance, a subscriber may not agree to accept a call from a particular person unless a lawyer is also included in the call. Thus, we cannot use the definition of composition to infer formally acceptability of goals from the acceptability of composite goals. However, in such situations we use heuristic rules to make *assumptions* about acceptability in the absence of explicit assignments for goals. One of these rules states that when a composite goal is neither acceptable nor unacceptable and all component goals are acceptable, then we assume that the composite goal is acceptable. Similarly, a second heuristic rule states that when a composite goal is neither acceptable nor unacceptable and at least one component goal is unacceptable, then we assume that the composite goal is unacceptable.

An acceptability marking of a goal hierarchy is made *complete* by repeatedly using these inference rules and heuristic rules, until application of the rules will not result in any more changes of markings. Completing the marking will, in general, diminish the searching needed for determining applicability of received proposals.

4.3 Outline of the negotiation process

A negotiation process consists of a number of separate tasks: specification of policies, generating proposals, determining acceptability of proposals, and generating counter-proposals. We illustrate using the example how negotiation would proceed using the definitions of the previous section.

Specification of policies. The task of the agents in our negotiation mechanism is to reach agreements that achieve short-term goals and meet long-term constraints,

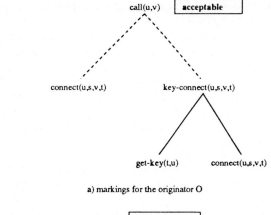

a) markings for the originator O

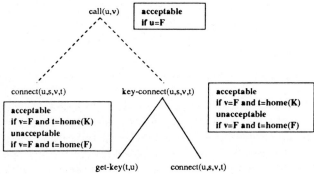

b) markings for forwarding party CF

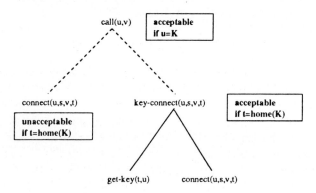

c) markings for K (subscriber to TKCP)

Figure 3: Acceptability markings.

or policies. In our example, the short-term goal is to set up a call. Long-term constraints are specified by policies. A policy for K (the user that subscribes to TKCP) is that her line cannot be accessed without entering the correct key. Policies can be specified separately from any particular negotiation process by marking an agent's goal hierarchy with the labels *acceptable* and *unacceptable*. Figure 3 illustrates such markings for three subscribers: O, the originator of a call, who will accept any call at all; F, a subscriber to Call Forwarding, who has forwarded calls to line L belonging to subscriber K; and subscriber K, a subscriber to Terminating Key Code Protection, who has set a key to screen calls from anyone not knowing the key.

Generating a proposal. The negotiation process is initiated when a goal is identified that needs to be achieved (e.g., when a subscriber indicates that she wants to make a call). The goal itself may not be acceptable, but, assuming that it is also not unacceptable, an acceptable specification of that goal could exist. The hierarchy is searched to find an acceptable specification of the goal. For example, any specification of call(O,F) is acceptable to O. Suppose that O's agent sends {call(O,F)} as the initial proposal.

Determining acceptability. When a proposal or counter-proposal is received by an agent, it has to decide whether it is acceptable. If it is, the agent can agree to it; if not, an alternative that is acceptable to the receiving agent needs to be generated. Acceptability is determined by searching the hierarchy. If the proposal is a specification of at least one acceptable goal, the proposal is acceptable. If it is the specification of at least one unacceptable goal, the proposal is clearly unacceptable. If none of the goals achieved by the proposal are acceptable or unacceptable, the proposal itself can still be acceptable. Intuitively, if a proposal can be subdivided into parts each of which achieves an acceptable goal, then the proposal is acceptable. Whenever a proposal is not acceptable, a counter-proposal is generated if there are any acceptable ones that have not already been explored. The following illustrates when proposals are acceptable.

1. The proposal {call(O,F)} is neither acceptable nor unacceptable to F. However, one of its specifications, namely {connect(O,home(O),F,home(K))}, is acceptable to F. Suppose that F sends this specification as a counterproposal.

2. The counter-proposal connect(O,home(O),F,home(K))} does not achieve a goal that is acceptable to K. But it achieves the goal {call(O,F)}, which is neither acceptable nor unacceptable to K (some of its specifications are acceptable, and some are not). K's agent can look for an alternative specification of this goal that will be acceptable to K.

3. The proposal {get-key(O, home(K)),connect(O,home(O),F,home(K))} is acceptable to K. It is also acceptable to O.

Generating a counter-proposal. When a proposal is not acceptable, we need to find an alternative. We distinguish three cases:

1. The proposal itself may not be acceptable, but sometimes a more detailed specification can be found that is. For instance, the proposal {call(O,F)} is not acceptable to F, but {connect(O,home(O),F,home(K))} is.

2. The proposal may achieve a goal that is unacceptable, but possibly also another, not unacceptable goal (e.g., an abstraction of that unacceptable goal) for which an acceptable specification can be found. The goal `connect(O, home(O),F,home(K))` is unacceptable to K, but its abstraction `call(O,F)` does have an acceptable specification.

3. If neither of these cases hold, we cannot find a solution without changing the acceptability markings in at least one of the agents. This would require relaxing the constraints of at least one of the involved agents. How to do this is an interesting question requiring further research.

The iterative generation of proposals and counter-proposals constitutes a search process through the goal hierarchy to find a specification that achieves the original goal of the agent that initiated the negotiation and is acceptable for all involved parties. The negotiation process terminates when either such a proposal is found, or one of the parties has exhausted all possibilities for generating counter-proposals.

5 Using negotiation on the Touring Machine platform

A negotiating system has been implemented on the Touring Machine platform [1]. This implementation depends on the operations provided by the Touring Machine platform, so we begin with a brief discussion of these operations.

These operations are applied to a Touring Machine structure called a *session*, which corresponds to the idea of a call in today's telecommunications systems. The basic structure used to define a session is called a *connector*. A connector contains a set of *sources*, which send information, and *sinks*, which receive information. Every sink receives from every source. The method used to combine signals from two or more different sources depends on the *type* of the connector. For example, audio signals are combined into one by a weighted addition. Video signals are combined into one by dividing the monitor screen into segments and displaying each signal in its own segment of the screen.

The operations provided by the Touring Machine platform for operating on sessions are:

addClient(x,y)	add clients x and y to a session
addCon(x,t)	add a connector named x of type t
delCon(x)	delete a connector named x
addSink(x,w,p)	add a sink to connector x, for user w, using logical device p
addSource(x,w,p)	add a source to connector x, for user w, using logical device p
setPrivacy(v)	set privacy mode v (who can know about the call) v
setPermission(v)	set permission mode v (who can operate on the call)

To initiate a two-party audio/video call between subscribers a and b, the following collection of operations would be used:

addClient(a,b)	specify the parties to the call
addCon(w, "audio")	
addSink (w, a, "speaker")	describe an audio connector from party b to party a
addSource (w, b, "mic")	
addCon(x, "audio")	
addSink (x, b, "speaker")	describe an audio connector from party a to party b
addSource (x, a, "mic")	
addCon (y, "video")	
addSink (y, a, "monitor")	describe a video connector from party b to party a
addSource (y, b, "camera")	
addCon(z, "video")	
addSink (z, b, "monitor")	describe a video connector from party a to party b
addSource (z, a, "camera")	
setPrivacy ("group")	describe who can be aware of the call
setPermission ("public")	describe who can join the call

In figure 4, we give the part of the goal hierarchy for the Touring Machine having to do with two-party calls. In this hierarchy, we provide for Call Forwarding and Voice Mail in addition to normal call setup.

To define a goal hierarchy, a provider uses a Prolog-like language. The definition contains a collection of rules, one for each composite goal and one for each specialization of an abstract goal. The rule for a composite goal has the form:

$$C \to S_1, \ldots, S_n$$

where C is the composite goal and $S_1, \ldots, S_n$ are its subgoals. The rule for a specialization of an abstract goal has the form

$$A \to S$$

where A is the abstract goal and S is its specialization.

A goal hierarchy will be most useful if the same one is used by all entities involved in the telecommunications system. Otherwise, an entity may receive a proposal involving goals that it doesn't recognize. If it always accepts such a proposal, it gives up the possibility of constraining the calls it receives. But if it always rejects such a proposal, it will be refusing calls that it might be willing to accept. In a more sophisticated system, it might be possible to query the originator of a proposal as to what it means, and if all subscribers recognize the same base set of operations on the platform, the answer could be given in the form of a new subhierarchy to be added to the original hierarchy. The subscriber would then have to supply the constraints for this new subhierarchy. At present, however, we do not do this, and for this reason, all Touring Machine subscribers use the same goal hierarchy.

To define constraints, a user or provider gives a statement about each possible state that he could be in. This statement specifies which instantiated goals are acceptable and which are not. It has the form

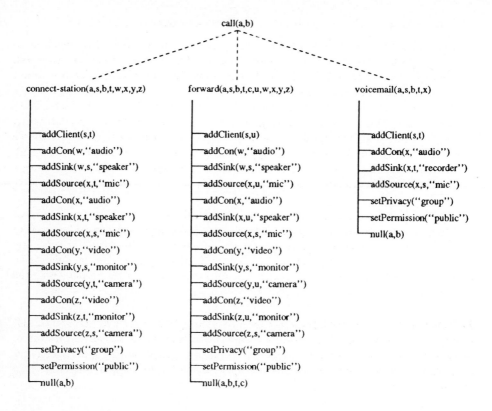

Figure 4: The goal hierarchy for two-party Touring Machine calls. Space limitations require the subgoals of the composite goals connect-station, forward, and voicemail to be written in a vertical column.

$$\text{when } S \text{ acceptable } G_1, \ldots, G_n \text{ unacceptable } U_1, \ldots, U_n$$

where S is the state name and the G_i and U_i are instantiated goals. (An instantiated goal is a goal whose arguments are all constants or the symbol ANY, meaning any value). Each entity prepares an individual collection of constraints, representing its own policy, based on the goals in the goal hierarchy.

The system entities currently represented by agents in the Touring Machine system are users and stations, or locations. At present, the system itself doesn't have an agent, because any proposal is acceptable to it. However, if there were restrictions on resource usage or if billing were included in the system an agent for the system would be useful.

Some issues that a designer of a telecommunications system incorporating a negotiation system must face are:

What are the user goals and how are they related? We have found it most intuitive to make $call(a_1, \ldots, a_n)$ a high-level goal (initiating a call among the specified parties), and then define as specializations various ways of handling the call, including forwarding, voice-mail, postponement, and break-in. However, we can also define even higher level goals than call, such as the goal to set up a set of calls all delivering the same message. This would be a composite goal using call repeatedly as a subgoal.

Which agent should authorize which operations? A variety of issues may arise in connection with this. For example, would a company agent authorize all calls to or from any employee agent, or would the employee agents have individual policies? Or would both the company and the employee agent be involved in each call to an employee?

How will the user interface be designed? One possibility is to provide a hierarchical system of menus, based on the goal hierarchy itself. A generic interface could be built that would work with any goal hierarchy. Another possibility is to use an existing interface, and relate requests from it to goals in the goal hierarchy to initiate negotiation. The Cruiser interface to the Touring Machine system has been used in our work [3].

6 Examples of AIN or CS-1 features

The description of the negotiation mechanism in section 4 contained examples of how the features Call Forwarding and Terminating Call Key Protection can be provided by our mechanism. Call screening features are also easily provided by defining appropriate acceptability predicates for parties to a call.

In this section, we illustrate how our mechanism can be used to provide several other complex features on the Touring Machine platform. These features are closely related to the IN features defined in the CCITT Q1200 IN standard [2, Q1211, pages 29–45], and the implementations described here can be used as a guide for how those IN features would be implemented using negotiation.

Some IN-like features have been implemented on top of the Touring Machine system, but not all IN features are meaningful in that environment because of the difference in functionality provided by the current public network or the Touring Machine system. For instance, Abbreviated Dialing is not meaningful in the Touring Machine system, since the 'dialing' interface is menu-based.

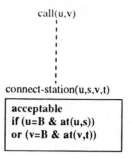

B's goal hierarchy

Figure 5: Goal hierarchy for Call Following. The Touring Machine parameters that specify connector names have been omitted.

6.1 Re-routing features

We describe in detail the Touring Machine feature *Call Following*. Although Call Following is not an IN feature, IN features such as Follow-Me Diversion and Destination Call Routing can be implemented similarly. This feature allows calls to a user to be redirected automatically to the station nearest to where the user is at that moment. The user's location is tracked by the Touring Machine name-server, which is notified whenever the user's locator badge moves to a different room. As long as a user wears the badge, her location will be known to the system.

The goal hierarchy that is used to implement this feature is similar to the one used for Call Forwarding (see Figure 5). The goal connect-station(B,S,V,T) is acceptable to user B in this example if and only if user B is at station S. Similary, the goal connect-station(U,S,B,T) is acceptable to user B if and only if user B is at station T. The predicate at(u,s) asserts that user u is at station s. We assume that a user's agent knows his or her location. If necessary, the agent can determine the location by querying the Touring Machine name-server.

Imagine the following situation. User A wants to call user B, who has Call Following (e.g., the hierarchy of Figure 5) and is currently visiting user C. If A's agent sends the proposal call(A,B), then B's agent will decide that the proposal is indeterminate and look for an acceptable specialization, i.e., a goal connect-station(A,S,B,T) such that the predicate at(B,T) is true. The agent can find such a station T by querying the name-server. This results in the counterproposal connect-station(A,S,B,T) where user B is at station T.

According to B's marking of the hierarchy, it doesn't matter what station is used for A. This may not be precisely true, but it is not reasonable to expect user B's agent to know where user A wants the call directed. Thus B's agent relies on A's agent to supply this information. A's agent may respond with a further counterproposal with yet another station substituted for S.

Alternatively, if A's agent sends the proposal connect-station(A, home(A), B, home(B)) (where home(A) and home(B) are A's and B's default stations), then B's agent will substitute the station at which B is currently located for home(B).

Follow-Me Diversion and Destination Call Routing can be implemented similarly. However, for those features an agent wouldn't need to query the name-server. Instead,

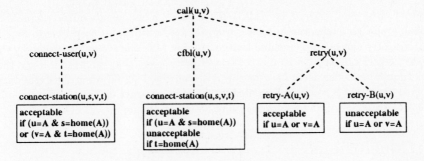

a) markings for A in idle state

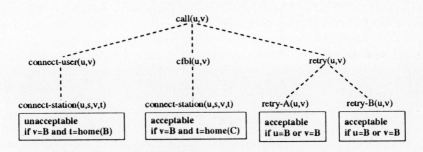

b) markings for B in busy state

Figure 6: Goal hierarchies for dealing with busy states.

its own data would indicate where to find the user.

6.2 Features that deal with busy states

Several features deal with busy states: Call Waiting (CW), Call Forwarding on Busy Line (CFBL), Completion of Calls to Busy Subscriber (CCBS), etc. Each of these features provides an alternative way of handling a call when the callee is busy and can be represented as different specializations of making a call. We show how such features can be implemented in our mechanism.

Figure 6 shows goal hierarchies for users A and B. As described in section 5, B has specified a marking of the goal hierarchy for the busy state, which is the marking shown here. The goals call(u,v) and connect-station(u,s,v,t) are defined as in the previous examples. The goal retry(u,v) denotes that call set up will retried later, either by the caller (retry-A) or by the callee (retry-B). The goal cfbl(u,v) implements CFBL. The markings place constraints on the way calls should be handled in certain states and represent a user's intentions regarding call handling.

In this example, we assume that user A tries to call B and that B is busy. User A is subscribed to the CCBS feature. B is subscribed to CFBL. Calls that are coming in for B when B is busy are re-routed to the station of user C.

A's agent sends call(A,B) as a first proposal. Since B is in the busy state, B's agent uses the hierarchy of figure 6 and finds that this proposal is indeterminate. Looking at specializations of call(A,B), the agent also finds that no specialization of connect-station(A,s,B,t) is acceptable to B. There are two possible counter-proposals: connect-station(A,s,B,home(C)) and retry(A,B). The first of these is acceptable to A and forward the call to C's station. The second counterproposal will result in a new counterproposal by A (`retry-A(A,B)`), which is acceptable to B.

A call retry can be implemented for the Touring Machine system by using database triggers. A user can define a temporary trigger in his agent that catches the event that a call is terminated. The agent then performs a predefined action when the event occurs. The goal `retry-A(A,B)` is implemented by defining a trigger on the termination of B's current call and specifying a corresponding action of setting up `call(A,B)`.

7 Conclusions

We claim that negotiation is more practical than conventional methods for solving the problem of policy feature interactions because it avoids the need for service providers or users to know what policy features other service providers or users have. This claim is based on the fact that conflicts between policies are settled at run-time by an automatic negotiation system that detects and resolves the conflicts (or interactions) at run-time. More conventional methods would examine pairs of features individually, and manually determine the resolution policy for each pair. Furthermore, negotiation considers the actual intentions of the user at run-time, so that individual and even changing policies are possible.

We have already established in earlier work that this kind of negotiation can be automated [4, 5]. This required defining an architecture for negotiation and developing algorithms to do the negotiation. These algorithms detect and resolve multi-party interactions at run-time. A single-party interaction can also be detected automatically, although the resolution is up to the user. An algorithm for completing an initial marking of goals in a goal hierarchy is described at the end of section 3 of [5]. This algorithm will also detect inconsistencies between constraints (an inconsistency is found if some goal is marked both acceptable and unacceptable by the algorithm).

In the present paper we have shown how to add a negotiating mechanism to a telecommunications platform. As long as the platform offers a well-defined set of operations on its interface, this mechanism will work. Furthermore, the properties of the negotiating system guarantee that some agreement will be found by the system if any exists that achieves the goal of the initiating subscriber [5].

Challenges that remain to be addressed include:

- How expressive can the language defining constraints be, before it becomes intractable to detect single-user feature interactions?

- How efficient is negotiation?

- Can constraints be used to guarantee safety properties?

- Can an automatic mechanism be devised for relaxing user constraints when no agreement can be reached otherwise?

References

[1] M. Arango, P. Bates, G. Gopal, N. Griffeth, G. Herman, T. Hickey, W. Leland, V. Mak, L. Ruston, M. Segal, M. Vecchi, A. Weinrib, and S.-Y. Wuu. Touring Machine: a software infrastructure to support multimedia communications. In

MULTIMEDIA '92, 4th IEEE COMSOC International Workshop on Multimedia Communications, Monterey, CA, 1-4 April 1992.

[2] CCITT. New Recommendations Q1200 – Q series: Intelligent Network Recommendation. Technical report, CCITT, COM XI-R 210-E, 1992.

[3] R.S. Fish. Cruiser: A multi-media system for social browsing. *ACM SIGGRAPH VIDEO REVIEW Supplement to Computer Graphics 45,* 6, Videotape, 1989.

[4] N.D. Griffeth and H. Velthuijsen. Reasoning about goals to resolve conflicts. In M. Huhns, M.P. Papazoglou, and G. Schlageter, editors, *Proceedings International Conference on Intelligent Cooperating Information Systems (ICICIS-93)*, pages 197–204, Rotterdam, May 12-14 1993. IEEE Computer Society Press.

[5] N.D. Griffeth and H. Velthuijsen. Win/win negotiation among autonomous agents. In *Proceedings 12th International Workshop on Distributed Artificial Intelligence*, pages 187–202, Hidden Valley, PA, May 19-21 1993.

[6] H. Rubin. An Introduction to the INA Field Experiment Initiative. Bellcore Special Report SR NWT-002280, June 1992.

Detecting Feature Interactions in the Intelligent Network

S.Tsang[1] and E.H.Magill
University of Strathclyde, UK
November, 1993

Abstract

This paper describes a distributed run-time approach to detecting feature interactions. The technique monitors feature and resource state transitions in run-time and by comparing these to "state signatures", which represent valid state transitions, it is possible to determine if a feature is operating correctly. It is assumed that incorrect operation is caused by feature interaction. The approach requires no centralised "a priori" knowledge of features nor any knowledge of how the features work, thus preserving service privacy. Encouraging preliminary results have been obtained from testing the approach in a software testbed designed during the project.

1. Introduction

The feature interaction problem [1,2] can be simply defined as the unwanted interference between two features. It represents an obstacle to the development and implementation of the Intelligent Network, undermining the principle that new features can be added rapidly. Existing solutions to the problem concentrate either at the specification stage [6, 7, 8, 9, 10, 11, 12] of the features' software lifecycle or while the features are actually running on the network [13, 14, 15, 16, 17, 18]. Although both approaches are strongly advocated, this paper concentrates on a run-time approach.

In the IN, features will be provided by different service providers and installed onto the network via the Service Management System (SMS). Therefore, the first time that features will encounter each other is when they are running together on the network. Furthermore, it is unlikely that competing service providers will disclose their feature specifications or even detailed information about the operation of their features to a central agency. Hence, solutions to this problem must act while the features are running, without employing centralised knowledge of the features. It is essential that solutions in the run-time domain cause as little service delay as possible in the network.

---

[1]This research is funded by the University of Strathclyde and the UK Science and Engineering Research Council.

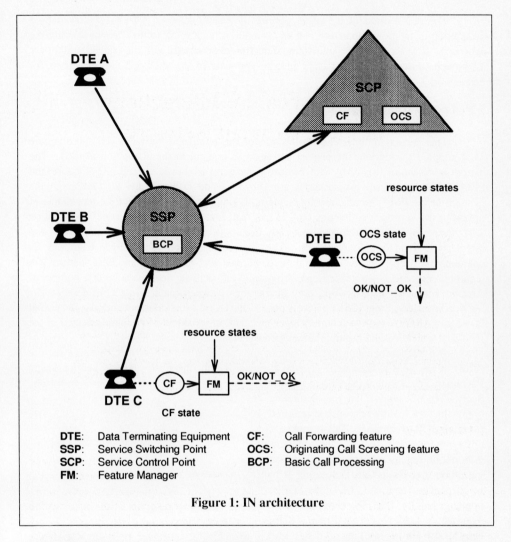

Figure 1: IN architecture

This paper describes a novel approach to detect feature interactions within the run-time domain which attempts to meet these demands. A Feature Management System (FMS) is proposed which, unlike some of the existing run-time techniques [14, 15, 16], *does not require centralised knowledge* about the features. Moreover, the approach does not require information on how the features work, so the features are *anonymous* to the FMS and *service privacy is preserved*. The approach can also detect interactions between *any number of features*. Because the techniques used within the FMS are simple, requiring no inter-agent negotiation or artificial intelligence techniques [17,18], the new approach is also *fast*. Furthermore, the FMS is a *low-cost* approach, requiring no modifications to the IN architecture. It was simulated and verified on a specially designed IN testbed which was implemented in software (using 'C' code) on a SUN workstation.

The first section of the paper defines the IN model used. This is followed by a description of the novel approach and an example of feature interaction detection using the approach. Results of feature interaction detection experiments with the testbed system are also given. Finally, the paper ends with an appraisal of the approach and directions for future work.

2. The Experimental Testbed

This section outlines the *abstraction* of the Intelligent Network (IN) architecture [3,4] adopted throughout the project and used to implement the simulation testbed. The overall architecture (together with Feature Managers) is shown in figure 1. Feature and Feature Manager (FM) instantiations are shown beside DTEs C and D. Although the location of the FM is an engineering concern, this matter does not affect the *mechanics* of the proposed approach so long as the FM can read feature and resource states (see section 3.3). In the testbed, FMs existed separately from both the Service Control Point (SCP) and the Service Switching Point (SSP). This logical separation of the FMs from the IN components meant the testbed could be developed and tested in a modular manner. A feature is regarded as a Service Logic Program (SLP) with services made up of a number of different features. In this work, only interactions between features are considered. The Service Switching Point (SSP) controls normal POTS calls and contains both the Basic Call Processing (BCP) and the feature trigger mechanism. A single SSP is implemented in the IN model which handles multiple simultaneous calls.

If a telephony event matches one of the entries in the trigger mechanism, a feature (SLP) in the SCP is activated. The SCP contains both the SLPs and the Service Logic Execution Environment (SLEE) to run the SLPs. When a feature has finished acting on a trigger, a feature response is returned from the SCP to the SSP. This response is different for each feature and contains information such as new digits or an indication that Basic Call Processing is to be suspended for a call. A single SCP is implemented in this IN model. Four features: Call Forwarding (CF), Originating Call Screening (OCS), Call Waiting (CW) and Three-Way Calling (3WC) are simulated in the testbed. The trigger mechanism governs the order in which features are instantiated and the order in which features receive telephony events. An arbitrary order of CF, OCS, CW and then 3WC is used by the trigger mechanism but this can be changed.

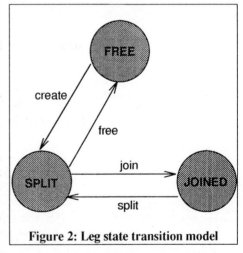

Figure 2: Leg state transition model

A user can access the network via a Data Terminating Equipment (DTE) such as a simple telephone receiver. The handset can be in either ONHOOK or OFFHOOK states and the tone generator can play standard tones such as DIAL_TONE and ENGAGED_TONE in addition to feature specific tones and announcements. Each DTE is connected to the switch (SSP) by a leg which can be in one of three states: FREE, SPLIT or JOINED.

The legs and DTE tone generators are the shared resources within the network and are manipulated by the BCP and features using Functional Component calls (FC calls) [5]. The set of FC calls available to manipulate the legs are: FREE, CREATE, SPLIT and JOIN. The state transition model of the legs is shown in figure 2.

3. The New Approach

3.1. Monitoring Feature and Resource States

A feature follows a strict execution sequence, determined by the Service Logic Program during which it responds to telephony events and manipulates network resources. This results in a sequence of state transitions which is always the same for a particular set of input telephony events. The states of the resources represent the result of the feature's actions, and by combining these with the feature state, a states combination is formed which reflects if a feature's actions were successful. This is shown in figure 3. For example, if the three-way calling feature is in the three-way call state and only two of the legs in the call are joined, it is clear that the feature has not operated correctly.

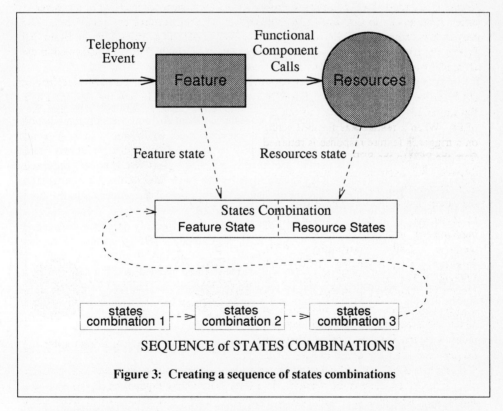

Figure 3: Creating a sequence of states combinations

Furthermore, if it is assumed that the feature is error-free and is running without any other concurrently active features, the state transitions recorded while it is running will represent the "correct" operation - or *"state signature"* - of the feature. It should be noted that these *state transitions are anonymous*, being represented as a sequence of digits.

A fault in the operation of the feature is detected when the state transitions do not match this "state signature". Based on this technique, the Feature Manager (FM) architecture shown in figure 4 was devised to monitor the operation of a single feature.

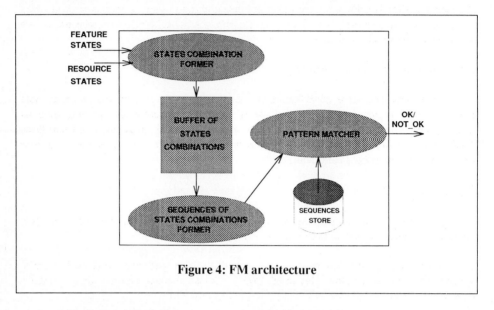

Figure 4: FM architecture

The Feature Manager (FM) performs only a few simple operations: the first is to read the feature and resource states and concatenate them to form a states combination. This combination is then added to a buffer of previous states combinations and a sequence of states combinations (the state transitions) is formed from the most recent states combinations. Finally, this current sequence is matched against the "correct" sequences held in the FM's own store - it is explained later in this paper (see section 3.2.2) how these are obtained. If the sequence cannot be matched against one of the sequences in the store, the FM reports this to the FM (using a NOT_OK signal).

It should be noted that all the FMs are the same. The only difference between an FM to monitor a call forwarding feature and an FM to monitor an originating call screening feature is the contents of the sequence store. Consequently, if an existing feature is updated, its associated sequence store will also have to be updated, and if a new feature is created, a new sequence store will be required to represent its "correct" sequences. This means that the same FM architecture and programming can be used to manage any existing or future feature.

3.2. Feature Management System

3.2.1. How It Works

When a new feature is instantiated, a Feature Manager (FM) instantiation, as shown in figure 4, is also created. The FM instantiation includes a copy of the appropriate sequence store which it needs to monitor the feature and does not intercommunicate with the other FMs. To implement this, the IN feature instantiation mechanism must be modified to additionally create a FM instantiation when it creates a new feature instantiation.

This mechanism must also ensure that the appropriate sequence store is included with the FM instantiation. The sequence stores reside in a central database. Once instantiated, the FMs do not intercommunicate, and each FM only monitors one feature instantiation. Because the FMs do not intercommunicate, their operation and performance is identical regardless of whether the interacting features belong to one or many subscribers, or are distributed across many switches.

3.2.2. Obtaining the Sequence Stores

The sequence store of a feature is the record of the feature and resource state transitions which occur while the feature is operating correctly and without external interferences (such as other active features or network failures). It is suggested that these conditions are met during trial runs of the feature on the IN and that a feature's sequence store should be obtained during this phase.

3.3. Assumptions

The new approach makes several assumptions about the information available to the Feature Management System (FMS) in the Intelligent Network:

1. It is assumed that the feature and resource states can be read by the FMs within a time span which causes acceptable delay to the FMs. It is also assumed that the communications overhead used to read the states is acceptable to the network.

2. The approach assumes that every time a feature acts on a telephony event, a state transition occurs, either in the feature or the resources the feature uses. Therefore, features which do not comply with this assumption will cause the FM to fail. It should be noted that this aspect of a feature - the number of states it uses - is implementation specific.

3. Furthermore, the FMS ignores data used by the features, so the approach may fail with features whose execution path depends on the value of data, such as digits dialled or time of day. An example of such a feature is originating call screening which acts differently depending on the number dialled by the subscriber. It is suggested that this effect can be minimised by employing a richer set of feature states and a comprehensive training phase. For example, it was possible to detect feature interactions with the Originating Call Screening (OCS) feature without monitoring the data because the feature, after analysing dialled digits, entered either a "call_screened" or "call_allowed" state. Clearly, if a call was set up (detected by leg state transitions) while the feature was in the "call_screened" state, an error had occurred.

4. Example Of Feature Interaction Detection

This section describes a feature interaction between the call waiting (CW) and three-way calling (3WC) features and how the new approach described in this paper, is used to detect the interaction. The result was obtained from a series of experiments with the simulated Intelligent Network (IN) testbed and Feature Management System (FMS). The section begins by describing the state transitions which occur while the features are operating correctly and then describing the state transitions when the features interact.

4.1. State Transitions of Features

The following state transitions represent each feature operating correctly and form part of the sequences store. Although the feature and resource states are shown here with names, it should be noted that the Feature Managers viewed these state transitions as *sequences of digits*.

4.1.1. Call Waiting

The call-waiting feature allows a subscriber who is already busy in an active call to receive calls from a third party. When a third party dials the subscriber, the subscriber hears a special call waiting tone (CW_tone) and can join the third party into the call by pressing the "recall" button on the handset. Note that the call-waiting feature does not create a three way call because it always leaves one of the parties split from the call.

Consider the following scenario: user A has subscribed to the call waiting feature. Initially, user A is involved in a normal POTS call with user B. Each party's DTE in the call is connected to a switch (SSP) by a leg. The *correct* feature and resource state transitions are:

CW feature state	A's leg state	B's leg state	C's leg state
= INACTIVE	= JOINED	= JOINED	= FREE
	A's DTE tone	B's DTE tone	C's DTE tone
	= none	= none	= none

If a user C then dials user A, user A's call waiting feature will activate and user A will hear a special call waiting tone. Moreover, user C will not hear the engaged tone but the normal ringing out tone:

CW feature state	A's leg state	B's leg state	C's leg state
= Two-way call, C dialled A	= JOINED	= JOINED	= SPLIT
	A's DTE tone	B's DTE tone	C's DTE tone
	= CW tone	= none	= ringing out

If user A then flashhooks, user B is split from the call and user C is joined:

CW feature state	A's leg state	B's leg state	C's leg state
= Two-way call, C joined in	= JOINED	= SPLIT	= JOINED
	A's DTE tone	B's DTE tone	C's DTE tone
	= none	= none	= none

and if user A flashhooks again, user B is rejoined into the call and user C split:

CW feature state	A's leg state	B's leg state	C's leg state
= Two-way call, B joined in	= JOINED	= JOINED	= SPLIT
	A's DTE tone	B's DTE tone	C's DTE tone
	= none	= none	= none

and so on until one of the parties onhooks and the feature returns to the INACTIVE state.

4.1.2. Three Way Calling

The three-way calling feature allows a subscriber, who is already involved in a POTS (two-way) call, to add a third party into the call. The feature is activated when the subscriber presses "recall" or flashhooks on the handset, splitting both parties from the call. The subscriber will hear a special dial tone as a cue to dial the number of the third party. If the dialled party answers, it is joined into the call with the subscriber, while the other party is split. If the subscriber presses "recall" again, a three-way call is formed.

Consider the following scenario: user A has subscribed to the three-way calling feature. Initially, user A is involved in a normal two-way POTS call with user B. Each party's DTE is connected to a switch (SSP) by a leg. The *correct* state transitions are:

3WC feature state	A's leg state	B's leg state	C's leg state
= INACTIVE	= JOINED	= JOINED	= FREE
A's DTE tone	B's DTE tone	C's DTE tone	
= none	= none	= none	

It should be noted that the 3WC Feature Manager will not be monitoring C's leg state because it is not involved in the call. After the first flashhook from user A, the 3WC feature is activated: user A hears the special dialling tone and can now dial a third party. Both user A and user B's legs are split from the call:

3WC feature state	A's leg state	B's leg state	C's leg state
= Wait for digits	= SPLIT	= SPLIT	= FREE
A's DTE tone	B's DTE tone	C's DTE tone	
= 3WC dial tone	= none	= none	

User A dials user C and the 3WC Feature Manager starts to also monitor C's leg state:

3WC feature state	A's leg state	B's leg state	C's leg state
= Waiting for answer	= SPLIT	= SPLIT	= SPLIT
A's DTE tone	B's DTE tone	C's DTE tone	
= ringing out	= none	= ringing in	

User C answers user A's call:

3WC feature state	A's leg state	B's leg state	C's leg state
= 3WC_AC_B	= JOINED	= SPLIT	= JOINED
A's DTE tone	B's DTE tone	C's DTE tone	
= none	= none	= none	

Finally, user A flashhooks to form the three way call:

3WC feature state	A's leg state	B's leg state	C's leg state
= 3WC_ABC	= JOINED	= JOINED	= JOINED
A's DTE tone	B's DTE tone	C's DTE tone	
= none	= none	= none	

4.2. State Transitions of Interacting Features

To demonstrate how the monitoring of feature and resource states can detect feature interactions, consider the following scenario: user A has subscribed to both the CW and 3WC features. Initially, A is involved in a normal two-way POTS call with user B when a third party (user C) dials user A, activating the CW feature. The resultant states are:

CW feature state = Two-way call, C dialled A	3WC feature state = INACTIVE	A's leg state = JOINED	A's DTE tone = CW tone
		B's leg state = JOINED	B's DTE tone = none
		C's leg state = SPLIT	C's DTE tone = ringing out

Note that the 3WC Feature Manager will not be monitoring C's states yet because it is not involved in the call yet - it will only start monitoring them after user A has dialled user C as the third party. At this point, both features are still working correctly. On hearing the CW tone user A flashhooks. Let us assume both features are offered the flashhook signal, but the CW feature acts first, resulting in the following states:

CW feature state = Two-way call, C joined in	3WC feature state = Wait for digits	A's leg state = SPLIT	A's DTE tone = 3WC dial tone
		B's leg state = SPLIT	B's DTE tone = none
		C's leg state = JOINED	C's DTE tone = none

Clearly, the CW feature is now not operating correctly compared with the state combinations and state transitions given in section 4.1.1 - A's leg should be in the JOINED state and there should by no tone - faults caused by the actions of the 3WC feature. If the CW feature was triggered after 3WC, the following are the resultant states:

CW feature state = Two-way call, C joined in	3WC feature state = Wait for digits	A's leg state = JOINED	A's DTE tone = none
		B's leg state = SPLIT	B's DTE tone = none
		C's leg state = JOINED	C's DTE tone = none

Now, the 3WC feature is not operating correctly when compared with the state combinations and state transitions given in section 4.1.2 - A's leg should be in the SPLIT state and A's DTE should be sounding the dialling tone - faults caused by the actions of the CW feature.

Four simulated features were implemented in the IN testbed: call forwarding (CF), originating call screening (OCS), call waiting (CW), and three-way calling (3WC). and combinations of these were used as test cases for the simulated Feature Management System (FMS). Experiments with the system were limited to interactions between two features and each feature combination could be implemented in one of two configurations:

1. both features sharing the same subscriber (single user configuration)
2. or each feature having a different subscriber (multiple user configuration)

The IN testbed is an interactive system in which the user can control the DTEs through a set of four telephony events: offhook, onhook, dial, and flashhook. The user also specifies which DTE generates the event and digits (if appropriate) with the selected event. In each test case, the goal was to find a feature interaction. This meant that the system had to be manipulated such that both features were simultaneously active. The results are given in section 5.

5. Results

The figure shown in figure 5 shows the results obtained from the experiments with the simulated Intelligent Network (IN) testbed using different feature combinations of four features: call forwarding (CF), originating call screening (OCS), call waiting (CW), and three-way calling (3WC). There were two possible configurations for each feature combination: either for both features to have the same subscriber (single user) or different subscribers (multiple users). Of the twelve feature combinations and configurations used, the FMS *failed to detect interaction* in one - call forwarding (CF) and originating call screening (OCS) with different subscribers - and reported interaction wrongly (a *false positive* result) for two - call waiting (CW) and originating call screening (OCS) with the same subscriber, and call forwarding (CF) and three-way calling (3WC) with the same subscriber.

The FMS failed to detect the interaction between CF and OCS because the CF feature was implemented in such a way that it did not cause any state transitions, violating the second assumption given in section 3.3. The false positive results were caused because feature X manipulated the resources shared with feature Y, resulting in resource states unrecognised by feature Y's FM. Consequently, feature Y's FM reported that an interaction had occurred. Although (strictly speaking) this is an interaction, the operation of feature Y was not affected by feature X's actions, so the reported interaction represented a false positive. The FMS did not recognise the distinction between a desired and a negative interaction.

6. Conclusions

This paper has outlined a run-time method for detecting feature interactions which is novel, simple and highly deterministic. An abstraction of the IN model has been used to create a testbed for the approach and experiments were carried out with combinations of four simulated features: call forwarding (CF), originating call screening (OCS), call waiting (CW) and three-way calling (3WC). The basis of the approach is to monitor sequences of feature and resource states for deviations from known "correct" state transitions (or state signatures) for the feature and announce these as feature interactions.

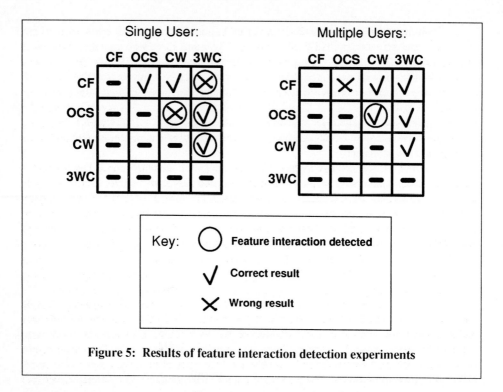

Figure 5: Results of feature interaction detection experiments

A Feature Management System (FMS) was outlined in which a Feature Manager (FM) instance, employing the states monitoring technique, is created for each feature instance. The FMs are all identical with only the sequence store used differing according to the feature. It should be stressed that the role of the FMs is not to manipulate the features, but to watch over them. The service providers may wish to take the responsibility of keeping the sequence stores up to date.

Because the approach requires no centralised knowledge of features or inter-agent communication, and requires only to read feature and resource states, service privacy is preserved. The functions performed by the FMs are simple, so the approach will also be fast, causing little additional service delay. Furthermore, the distributed architecture of the approach, with each FM managing a single feature, will allow it to detect interactions between any number of features. Because the FMs are autonomous and do not intercommunicate, their operation and performance is the same regardless of the number of users or geographical distribution of the interacting features. This means that the approach is potentially applicable to a very wide range of interaction problems.

While experimenting with the IN testbed, the problem of "transient" state combinations occurred when feature or resource state transitions completed at different times. This caused the number of sequences which had to be "learned" by the FMs to become very high. To resolve this, the FM interface which accessed the state information was "buffered" so that states which lasted less than (an experimentally chosen) 1 second were ignored.

Another potential problem is the assumption that the feature and resource states can be easily accessed, and in reality this may not be the case. However, this problem is implementation specific to such a degree that it lies out of the scope of this project.

Out of the twelve feature combinations used in the experiments, the FMS failed to detect interaction in one (between CF and OCS) and reported interaction wrongly in two (CF and 3WC, CW and OCS). The problem was that the approach was too deterministic and the slightest deviation from the state signature resulted in the FM reporting feature interaction. Even so, it should be noted that these are only preliminary results and more experimentation is planned.

The new approach, has many advantages over the existing approaches and initial results have been promising. It has been demonstrated that it is possible, using the simple techniques described in this paper, to detect certain feature interactions in the run-time domain, without needing centralised knowledge about the features in the network or resorting to complex artificial intelligence techniques [13, 17, 18]. Although it appears to be a disadvantage to employ only state information in the technique, this approach allows the size and number of sequences to be kept to a manageable level. The inclusion of data would cause a "state explosion". Furthermore, it has also been shown (see point 3, section 3.3) that by employing a rich set of feature states, it is unnecessary to monitor the data.

This paper does not claim that the new approach can detect every kind of feature interaction (see section 3.3) but it has been demonstrated using four features in single-user and multi-user configurations, and further work will involve testing the approach with more simulated features. Ongoing work will also investigate methods to "soften" the determinism of the approach and thus resolve the problem of detecting false positives (see section 5). Moreover, the actions which should be taken once an interaction has been detected have not yet been considered in depth and will also form future work.

References

[1] E.Jane Cameron and Hugo Velthuijsen, "Feature Interactions in Telecommunications Systems", IEEE Communications, August 1993, pp18-23

[2] Nancy D.Griffeth and Yow-Jian Lin, "The Feature-Interaction Problem", Computer, August 1993, pp14-18

[3] Roger K.Berman and John H.Brewster, "Perspectives on the AIN Architecture", IEEE Communications, February 1992, pp27-32

[4] Jose M.Duran and John Visser, "International Standards for Intelligent Networks", IEEE Communications, February 1992, pp34-42

[5] Helen A.Bauer, John J.Kulzer, Edward G.Sable, "Designing Service-Independent Capabilities for Intelligent Networks", IEEE Communications, December 1988, pp31-41

[6] R.A.Verstraete and H.J.M.Decuypere, "A Strategy for Studying Services and Service Interactions in Intelligent Networks", 1990 IEEE Globecom, pp1650-1654

[7] F.S.Dworak, "Approaches to Detecting Feature Interactions", 1991 IEEE Globecom, pp1371-1377

[8] Y.Harada, Y.Hirakawa, T.Takenaka, "A Design Support Method for Telecommunications Service Interactions", 1991 IEEE Globecom, pp1661-1666

[9] A.Lee, "Formal Specification - A Key to Service Interaction Analysis", SETSS 1992, 30th March - 4th April, pp62-66

[10] Rezki Boumezbeur and Luigi Logrippo, "Specifying Telephone Systems in LOTOS", IEEE Communications, August 1993, pp38-45

[11] Eric Kuisch, Ronald Janmaat, Harm Mulder, Iko Keesmaat, "A Practical Approach to Service Interactions", IEEE Communications, August 1993, pp24-31

[12] Yasushi Wakahara, Masanobu Fujioka, Hiroyuki Kikuta, Hikaru Yagi, Sei-ichiro Sakai, "A Method for Detecting Service Interactions", IEEE Communications, August 1993, pp32-37

[13] Nancy D.Griffeth and Hugo Velthuijsen, "The Negotiating Agent Model for Rapid Feature Development", SETSS 1992, 30th March - 4th April, pp67-71

[14] S.Homayoon and H.Singh, "Methods of Addressing the Interactions of Intelligent Network Services with Embedded Switch Services", IEEE Communications, December 1988, p42 ff

[15] Michael Cain, "Managing Run-Time Interactions Between Call-Processing Features", IEEE Communications, February 1992, pp44-50

[16] J.Aitken, "Feature Interaction in Mainstream IN Applications", First International Workshop on Feature Interactions, December 3-4, 1992, pp22-40

[17] M.R.Kendall, "An Artificial Intelligence Approach to Call Processing", BT Technology Journal, October 1990, pp30-40

[18] R.Weihmayer and R.Brandau, "Co-operative Distributed Problem Solving for Communications Network Management", Computer Communications, November 1990, pp547-557

Restructuring the Problem of Feature Interaction:

Has the Approach been Validated?

Experience with an Advanced Telecommunications Application for Personal Mobility

MARILYN CROSS and FERGUS O'BRIEN
Telecommunications Software Research Centre
University of Wollongong
Northfields Avenue
WOLLONGONG
NSW 2522
AUSTRALIA

mcross@cs.uow.edu.au
fob@s.uow.edu.au

Abstract

In an earlier paper the position was taken that one solution to feature interaction was to restructure the problem, so that the entities being added were not features, but non-interacting applications. The problem of feature interaction was recast as a realisation of the more abstract patterning of communication from the perspective of social interaction.

The problem was contextualised by taking the advanced service UPT (Universal Personal Telecommunications Service) as an example of a service that may be viewed as a means for social interaction and which has the potential for feature combination. One possible feature combination in that service was the well discussed round robin of *retry on busy* and *return call if busy*. It was suggested that the features in the service be recast as components of communication events which form part of a cohesive communication unit. By recasting the problem in such a way, the new paradigm afforded an organised range of parameters which included choice in the temporal connections between communication events, choice in the media by which events are conveyed and choice in the participants. What is seminal in both the description of the feature combination and in the recasting of the problem is the availability of the service user, where that availability may be defined in terms of the parameters of location, time and media.

In the UPT service defined as social communication, the parameters enabled the communication supported by the service to be asynchronous. If one varied the temporal parameter, the return call does not have to be immediate, it may be delayed a minute, an hour, a day depending on the exigencies of the situation. Extending the paradigm to what people do in such a situation, the calling party may well elect to call some time later. Alternatively, other asynchronous media are available, for example, communication may take place via a fax or electronic mail. A theoretical stance derived from a model of communication as social semiotic was suggested for the restructuring of the problem of feature interaction.

In order to validate the approach, the design and prototyping of an advanced telecommunications application that utilises UPT has been carried out. In that application, a system called Morgan for organising meetings using a range of media, it is the availability of the user that provides the pivot for the system. The Morgan system is able to manipulate a user's availability in location, time and media. The parameters of time, location and media are controlled at a 'metalevel' by means of user diaries that detail the availability of users, their location at any particular time and the type of media by means of which they are available.

Has the application validated the proposition that the kinds of feature interaction possible in the service are subsumed by the restructuring of the problem? It *has* in the sense, that the feature interaction is no longer problematic when the system is represented at a different level. At that level where the user's availability is central, the ability to manipulate availability events via the parameters of location, time and media has provided alternatives to handling the feature interaction problem described. In other words it need never occur. It *hasn't* in the sense that at a lower level in the representation of the system, the feature interaction may still occur.

The same paradox of appearing and disappearing feature interaction was also encountered in the object-oriented design and prototyping of the Morgan system where the features that were interacting were the result of low level interactions between data structures and methods ('functions'). The detection and solution of this feature interaction was to represent the structures in a matrix composed of data structures versus the methods to be performed on those structures. One example was that at a high level of data storage definition for persistent objects, there was no undesirable feature interaction. However at a deep level of storage there was interaction in the storage of circular versus simple lists. In the overall matrix of methods against data structures, it was possible to identify the paths through the matrix that resulted in feature interaction, and those that did not. The resolution of this interaction was to select another storage method, in this case the simple store method, that did not interact with the circular list data structure.

Analogously, telecommunications features may also be and are commonly represented as objects. Then it is also possible to propose the same matrix solution. This suggests one solution to the problem of detecting feature interaction, but the task still remains of linking the detection of the features at the deep level to the high level of the system where the feature interaction is no longer visible. It is concluded that feature interaction requires detection and solution at the deep level but is avoided at the highest level by re-representation, for example, in terms of the user's availability or lack of availability.

1. Introduction

In [1] the position was taken that one solution to feature interaction was to restructure the problem, so that the entities being added were not features, but non-interacting applications. The problem of feature interaction was recast as a realisation of the more abstract patterning of communication from the perspective of social interaction.

The problem was contextualised by taking the advanced service UPT (Universal Personal Telecommunications Service) as an example of a service that may be viewed as a means for social interaction and which has the potential for feature combination [2]. One possible feature combination in that service was the well discussed round robin of *retry on busy* and *return call if busy* [3]. It was suggested that the features in the service be recast as components of communication events which form part of a cohesive communication unit. By recasting the problem in such a way, the new paradigm afforded an organised range of parameters which included choice in the temporal connections between communication events, choice in the media by which events are conveyed and choice in the participants. What is seminal in both the description of the feature combination and in the recasting of the problem is the availability of the service user, where that availability may be defined in terms of the parameters of location, time and media.

In the UPT service defined as social communication, the parameters enabled the communication supported by the service to be asynchronous. If one varied the temporal parameter, the return call does not have to be immediate, it may be delayed a minute, an hour, a day depending on the exigencies of the situation. Extending the paradigm to what people do in such a situation, A may well elect to call B some time later. Alternatively, other asynchronous media are available, for example, communication may take place via a fax or electronic mail. A theoretical stance derived from a model of communication as social semiotic was suggested for the restructuring of the problem of feature interaction.

In order to validate the approach, the design and prototyping of an advanced telecommunications application that utilises UPT has been carried out. In that application, a system called Morgan for organising meetings using a range of media, it is the availability of the user that provides the pivot for the system [4]. The Morgan system is able to manipulate a user's availability in location, time and media. The parameters of time, location and media are controlled at a 'metalevel' by means of user diaries that detail the availability of users, their location at any particular time and the type of media by means of which they are available. An overview of the system is presented in section 5.

Has the application validated the proposition that the kinds of feature interaction possible in the service are subsumed by the restructuring of the problem? It *has* in Herbert's [5] sense, in that the feature interaction is no longer problematic when the system is represented at a different level. At that level where the user's availability is central, the ability to manipulate availability events via the parameters of location, time and media has provided alternatives to handling the feature interaction problem described. In other words it need never occur. It *hasn't* in the sense that at a lower level in the representation of the system, the feature interaction may still occur.

The same paradox of appearing and disappearing feature interaction was also encountered in the object-oriented design and prototyping of the Morgan system where the features that were interacting were the result of low level interactions between data structures and methods ('functions'). The detection and solution of this feature interaction was to represent the structures in a matrix composed of data structures versus the methods to be performed on those structures. This topic is discussed in section 6.

2. Feature Interaction

Features may be defined as packages of incrementally added functionality which provide services to subscribers or the telephone administration. Feature interaction

occurs when one feature interferes with the operation of another feature, or more specifically when:
> one feature inhibits or subverts the intended execution of another feature or another instance of itself, or creates joint execution dilemmas that are not resolved within the individual feature instances
> [6].

The exploration of feature interaction taken in the research reported here, most nearly fits into the level of feature interaction specified through logical relationships, where features interact because of logical relationships between the functions that they provide [6]. At the logical level, Bowen et al [6] have defined the feature interaction problem as that of determining how to specify features at the logical level so that undesirable feature interactions do not occur but useful feature interactions are permitted and the network specification can be checked easily for interactions.

Using the paradigm of communication as social interaction, the definition of features as packages of functionality, may be recast with features defined as cohesive packages of communication events. Not only are these events interrelated, but the parameters that identify such events may be delimited and theoretically justified. In contrast to other domains which are entity rich and may be captured in database systems (cf. [3]), telecommunication services and their component features are entity poor, but event rich. The representation of those features and the architectural environments which support such representations must attempt to capture such event rich patterns.

As the service application on which the current research is focused is Universal Personal Telecommunication, the next section briefly defines UPT and in particular the mobility that characterises the service. The UPT service is then recontextualised in a model of (social) communication (refer section 4) that then forms the basis for the architecture of the MORGAN application (refer section 5).

3. UPT and Mobility

UPT is a service that provides the user with personal mobility, offering the user telecommunication services irrespective of geographical location. The user is no longer indirectly identified by use of a particular terminal, for example, a specific phone number, but by a personal identifier - a network-transparent personal number called the UPT number [7].

3.1. Definition of UPT

The standards body CCITT [2], define UPT as a service which enables access to telecommunication services while allowing personal mobility. UPT enables each UPT user to participate in a user-defined set of prescribed services and to initiate and receive calls on the basis of a unique personal number, or identifier. The identifier is network-independent across multiple networks at any terminal, fixed, movable or mobile, irrespective of geographical location and limited only by terminal and network capabilities and restrictions imposed by the network provider.

3.2. Mobility in UPT

As may be seen from the previous section, mobility in UPT has been defined as geographic mobility. However, as O'Brien [8] has pointed out the model of personal mobility in UPT needs augmentation to incorporate temporal effects. In any attempt at communication, especially when (two) people are geographically dispersed, there may be only a relatively small percentage of time during the day when the parties are able to communicate. If UPT is extended towards Negroponte's vision [9] of the electronic surrogate with integrated telecommunications and computer facilities, then the choice of different communication media becomes available, for example, fax and electronic mail, then mobility is also extended to media. Thus, not only is geographic mobility important in UPT, but also mobility in time and media [10].

As mobility is such a critical parameter for UPT, it is worth examining the meaning of the term. Mobility implies movement within a range of options. With UPT one is not tied to a location, but has a choice (not unlimited) of a range of locations from where to communicate. Similarly, one is not limited to a particular time to communicate but has a choice of when to communicate. Choosing one particular time eliminates the other possibilities. Furthermore, there is a choice in the way in which one communicates. One may select to communicate using different media. The choice of the telephone commits one to a verbal interactive form of communication, whereas the choice of electronic mail, which is normally used in an asynchronous mode, commits one to a written non-interactive form of communication.

4. A Communication View of Telecommunication Services

The structure of communication reveals the collaborative nature of the activities that compose day to day interaction. Communication events are related in systematic ways: one communication event predicates the next in sequential and interactionally consequential ways, for example, a telephone call in which one leaves a message with a surrogate is likely to be followed by a return call from the targeted party.

In fact, communication in organisations, using a range of media, paces the business of the day. To paraphrase Boden [11] organisational members, their clients and suppliers spend a considerable amount of interactional energy coordinating activities in time and space while, at the same time, their very communication is itself a microcosm of that synchrony.

Telecommunications services are facilitators of the communication process in organisations. As such, they are most useful when they attempt to accommodate the collaborative interaction that characterises the communicative patterns of an organisation. Any patterning of services would do well to reflect the patterning of communication and the parameters that shape and guide that communication. It may be hypothesised that any decomposition into features is serving the more abstract functions of communication.

4.1. Communication Units and Events

An example of a unit of interaction is the scheduling for a meeting which is complete when everybody who will attend the meeting has responded in some way and the meeting has been scheduled. The unit of scheduling the meeting may be composed of a number of communication events, for example, a conversation to select alternative times and places, then a series of telephone calls to organise the attendees and negotiate a final time and venue, each of which forms a communication event.

The next section describes an advanced application of UPT (MORGAN) that organises meetings for UPT users capitalising on the mobility in location and media that the service offers and in addition utilising both these parameters and time to effectively schedule meetings and avoid the kind of feature interaction referred to previously.

5. Architecture of MORGAN

It is possible to put forward an object-oriented design of the communication model which represents the basic requirements for an advanced application of UPT such as organising meetings. This is given in figure 1. The UPT profile is part of the recommended standard [12]. The diary is a requirement for advanced UPT that extends the temporal dimensions of the service and stores the UPT user's availability in both time and media.

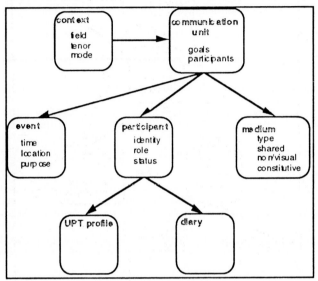

Figure 1: Object-Oriented Model of Communication for Advanced UPT

The collaboration between object classes is given in figure 2. It will be noted that the additional objects of *event* and *plan* have been introduced into the model to capture the dynamic nature of the *communication unit* which unfolds dynamically over time. In the MORGAN system additional objects include the MeetingOrganiser which incorporates the functionality for organising meetings and the KnowledgeBase which contains the rules by which the meetings are scheduled [4].

While the MORGAN system avoided the feature interaction problem by manipulating the temporal, location and media parameters, it was still possible to encounter feature interactions problems at a lower level in the system that required resolution. In the next section an example of that lower level of feature interaction is presented and a resolution suggested.

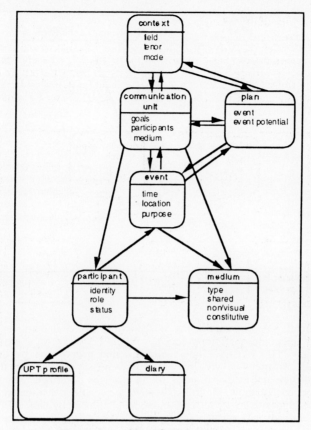

Figure 2:   Class Collaboration

6.   Resolution of Feature Interaction Using a Matrix

An example of feature interaction encountered at the lower level of the MORGAN system was one involving data storage and methods for manipulating that data. At a high level of data storage definition for persistent objects there was no undesirable feature interaction. However at a deep level of storage there was interaction in the storage of circular versus simple lists. In the overall matrix of methods against data structures, it was possible to identify the paths through the matrix that resulted in feature interaction, and those that did not.

In figure 3 this example is shown where two features are mapped on the method structure matrix by two separate paths, path 1 and path 2. A feature interaction is denoted by the circle on figure 3 where a deep store method is applied to a circular list data structure which leads to an infinite processing loop. The resolution of this interaction is to select another storage method, in this case the simple store method, that does not interact with the circular list data structure. The resulting feature map for path 2 is now replaced by path 3 so that the overall service shows no feature interaction.

Analogously, telecommunications features may also be and are commonly represented as objects. Then it is also possible to propose the same matrix solution in which the structures are represented in a matrix composed of data structures versus the methods to be performed on those structures. This suggests one solution to the problem of detecting and resolving feature interaction at the deep level, but the task still remains of linking the detection of the features at the deep level to the high level of the system where the feature interaction is no longer visible.

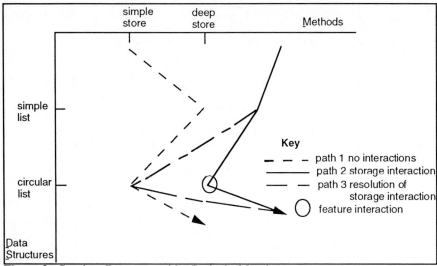
Figure 3: Services Represented by Paths in Matrix

7. Summary

Previously we proposed that it was possible to avoid feature interaction by restructuring telecommunications services in the context of a model of social communication [1]. We have now validated that proposition by designing and developing an advanced service in which it is possible to utilise the additional parameters of time and media in order to eliminate the described feature interaction. From our experience with the MORGAN system, we conclude that feature interaction is avoided at the highest level by re-representation, simplistically glossed in terms of the user's availability or lack of availability. However, it is still possible to encounter feature interaction at the deep level, for example, interactions between data structures and methods, that requires detection and solution. A matrix solution is suggested for resolution of the feature interaction at that deep level. The structures are represented in a matrix composed of data structures versus the methods to be performed on those structures. Both the detection and resolution of the interaction is given by designated paths through the matrix.

Our experience with validating the restructuring approach to feature interaction bears out Herbert's view [5] that feature interaction needs to be viewed at a number of levels in a system, starting at the highest level with the customer/ enterprise view of the system and descending through various levels. What is feature interaction at one level is not necessarily feature interaction at another. One of the next questions to explore is whether feature interaction detected at one level has or requires a representation at a different level and if so what is the relationship between the two.

References

[1] Cross M. and O'Brien F. 1992. Restructuring the Problem of Feature Interaction: A Perspective from Communication as Social Interaction. *Proceedings International Workshop on Feature Interactions in Telecommunications Software Systems*, Florida, December, 1992, 160-161.

[2] CCITT 1991. *Draft Recommendation F.851 - Universal Personal Telecommunications Service, Principles and Operational Provisions, Version 5.*

[3] Bowen T. F., Dworak F. S., Chow C. H., Griffeth N. D., Herman G. E., and Lin Y.-J., 1989. The Feature Interaction Problem in Telecommunication Systems. *Proceedings of the Seventh IEE International Conference on Software Engineering for Telecommunications Systems.*

[4] Bendeich J., Black I., James C., Ladmore J., and Sgangarella P. 1993. *Advanced Telecommunications Services: MORGAN*. User and Technical Manuals, Wollongong University, Department of Computer Science.
[5] Herbert A. 1993. Personal Comment at the Feature Interaction Session. *TINA '93*, L'Aquila, Rome, September 27 - 30.
[6] Bowen T. F., Chow C., Dworak F. S., Griffeth N. and Lin Y. 1990. *Views on the Feature Interaction Problem*. Manuscript. Morristown: Bellcore.
[7] Wilbur-ham M. 1992. UPT Studies in the CCITT. *Proceedings Telecom: Tokyo Forum '92*.
[8] O'Brien F., 1992. A User Negotiation Approach to UPT. *Proceedings of Telecommunications Information Networking Architecture Workshop*, Japan, January, 1992.
[9] Negroponte N. P. 1991. Products and Services for Computer Networks. *Scientific American*, 265, 76-83.
[10] Cross M. and O'Brien F. 1993. Advanced Personal Telecommunications Services: An Exploration of the Dimensions of UPT Using a Model of Communication - Models and Implementation. *Proceedings Telecommunications Information Networking Architecture Workshop (TINA93), Vol.II*, L'Aquila, Italy, September, 1933, 415-426.
[11] Boden D. 1990. People are Talking: Conversation Analysis and Symbolic Interaction. In H. S. Becker and M. McCall (eds) *Symbolic Interaction and Cultural Studies*. Chicago: University of Chicago Press, 244-273.
[12] CCITT, 1992. *Draft Recommendation F.851 - Universal Personal Telecommunications Service. Service Description, Version 8*

An Architecture for Defining Features and Exploring Interactions[1]

Douglas D. Dankel II
Mark Schmalz
Wayne Walker

Karsten Nielsen
Luiz Muzzi
David Rhodes

E301 CSE, C.I.S., University of Florida, Gainesville, FL 32611, USA

Abstract. The last decade has seen an explosive growth in the development of telephony features. The description and design of new features are fraught with errors due to this growth's impact on our ability to recognize interactions and the current practice of describing a feature's requirements using natural language. While the use of natural language eases the communication of requirements between the designer, customer, and developer, it introduces the potentially fatal flaw of ambiguity. Additionally, these requirements documents are rarely updated to reflect interactions with newly developed features. This paper presents an overview of a natural language-based system currently in development that converts English-based telephony requirements into a knowledge-based representation. The goals of this conversion are to create an unambiguous understanding of the requirements of the described telephony feature, to create less ambiguous written requirements documents, to automatically update existing requirements documents when new interactions are found, and to assist the feature designer in locating potential interactions with other features.

1. Introduction

Since the appearance of electronic switching systems (ESS) in the mid-1960s, computer technology has become the driving force in the development of telephony features such as automatic long-distance billing and three-way calling. Installed in 1965, the first No. 1 ESS provided 187 features, eleven of which were not available on the No. 5 Crossbar system which the No. 1 ESS replaced [1]. By the late 1970s the number of features had risen to more that 500 while approximately 5000 features are available today.

As the number of features increased, the burden of providing these features shifted from the hardware to the software. During the 1980s, the hardware component of a typical 15,000 line telephone office decreased by 45% (from 55 bays of hardware to 30) while the software component increased 1500% (from five feature packages to 80) [2]. The software library for these feature packages has grown from approximately 0.5 million

---

[1] This research is supported through the University of Florida/Purdue University SERC (Software Engineering Research Center) by an enhancement grant from BNR, Research Triangle Park, NC.

lines of code in 1979 to over 12.5 million today, making this features library one of the largest software libraries in the world [2].

The current software development/modification process for new features relies heavily on a designer's knowledge about the structure and organization of existing features. It is the designer's responsibility to determine what potential and actual interactions with other features exist and must be accounted for since the existing feature description documentation does not accurately reflect the implementation. With the growth in the number of available features, feature development has become highly problematic, placing the telecommunication industry at a crossroad. It can either develop tools that assist in managing this complexity thereby aiding a feature designer in determining potential feature interactions, or it can continue to use the existing process and see a steady decline in the productivity of employees.

This paper presents a model for a requirements capturing system (called GATOR), devised to assist designers in specifying the requirements of a new telephony feature and locating potential interactions with other features. First, in Section 2 we present a short summary of some of the causes for feature interactions. Section 3 discuss the problems of requirements capturing and our model of this process. Section 4 concentrate on POND, our knowledge representation language that is currently under development, which facilitates the encoding of features and interactions between features. In section 5 we return to the issue of feature interactions and discuss how POND can eventually assist the feature designer in identifying potential interactions. Finally, we report on the status of our system's development.

2. Feature Interactions Causes

Interactions between two features can occur in one of three ways. First, features can be *compatible with no specified restrictions*. For example, consider Speed Calling (SC) and Three-way Calling (TWC). When dialing a third party's number to link them into an existing conversation SC can be used since there are no compatibility restrictions between these features. The use of SC is clear and unambiguous. Second, features can be *compatible with special restrictions*. This form of interaction can be illustrated with Calling Number Delivery Blocking (CNDB) and Customer Originated Trace (COT). CNDB states that the calling party's number will not be delivered to the called party while COT states that the date and time of all calls to a number will be stored with the calling directory number (DN). Should both features be present, COT takes precedence, allowing the trace to be completed even though the CNDB customer's DN is not revealed to the called party. Finally, features can be *incompatible* either due to designated restrictions (e.g., Coin Lines are prohibited from allowing TWC) or due to conflicts between the features (e.g., the initiator of TWC is not allowed to receive Call Waiting calls or the Call Waiting tone since a hookswitch flash by the initiator would be ambiguous).

The latter two feature interactions (compatible with special restrictions and incompatible) result from a number of causes including:

1. Temporal conflicts resulting when actions specified within features are to occur at the same time. Both features require an action to occur at the same time, resulting in one feature taking temporal precedence over the other.
2. User input ambiguity resulting when features require the user to provide the same input to initiate different system responses. Because the same user response is required, this results in one feature being inhibited.
3. Resource conflicts occurring when features simultaneously require the same resource (which can also be classified as a temporal conflict) or when

one feature fails to provide a resource required by another. For example, Selective Call Rejection (SCR) and Automatic Recall (AR) have a resource conflict involving the calling party's number. The specification of SCR identifies that it allows a customer to block all calls originating from a particular number. This implies that calls from that number are never connected to the customer's line (e.g., the incoming memory slot associated with the line is not updated with the number) so this customer cannot perform an AR.

4. System response conflicts, which occur when features respond to the same situation. For example, when two features present on the same call both specify responses to the same situation, one of these actions should be given explicit precedence over the other.

5. Ambiguities and conflicts resulting from the distribution of software components. The introduction of the Advanced Intelligent Network (AIN) resulted in a number of these interactions between AIN features and ISDN and CLASS features [3].

The wealth of features and diversity of potential interactions present a serious problem to the specifiers and developers of new features. In the next section we briefly discuss requirements capturing and its importance in developing new features and introduce GATOR (the GATherer Of Requirements), a requirements capturing system designed to assist feature developers create more accurate and complete specifications.

3. Requirements Capturing and GATOR

Two characteristics confound accurate requirements development for telecommunications software. First, telecommunication software is large and complex. This software must perform real-time, event-driven processing -- processing so complex that it represents less than 1% of all computer software currently on the market [2]. This complexity impacts the development of new software since the designer of a new feature must be aware of the characteristics of thousands of features so any interactions can be accurately and completely described. Once described, the new feature must be implemented so it does not degrade the performance of the existing system. The second confounding characteristic is the medium currently used for describing features. All features are described using natural language (e.g., English), which can be highly ambiguous. The ambiguity within a requirements document often leads to customers and suppliers misinterpreting the requirements, which severely affecting the feature's implementation and acceptance.

Attempts to remedy this problem have involved the creation of software development environments [4], software information systems [5, 6], requirements capturing systems [7, 8], and specification languages [9, 10, 11]. These efforts typically have involved the use of a formal requirements description language to capture descriptions of new systems to be developed. Few have examined the problems of building a description of a complex existing system that requires modification and extension. GATOR [12], the GATherer Of Requirements, is a natural language based system designed to assist in the specification of functional requirements. Through its use of a natural language front-end, it attempts to address the issue of capturing and representing knowledge about the specifications of an existing system and using these captured specifications to assist in the development of new system features. GATOR consists of three major components (see Figure 1):

1. *The User Interface* consisting of GRoWl (the Graphical Requirements capturing Window interface), the Parser, the Lexical and Grammatical Knowledge Base, the Predicate Generator, and the Response Generator.

This interface accepts input from a feature designer in the form of natural language sentences and menu selections that describe or request information pertaining to new or existing telephony features. Natural language input is parsed by the Parser and interpreted by the Predicate Generator, while menu selections are converted directly to predicates because of their lack of ambiguity.

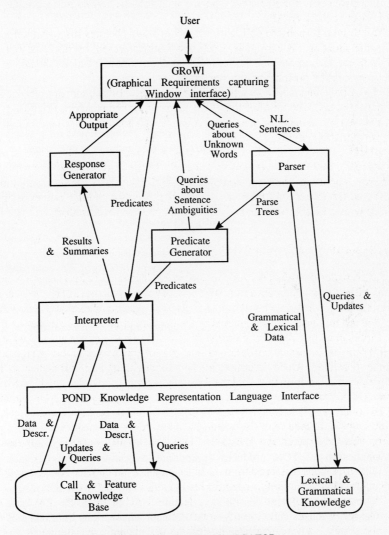

Figure 1. The Architecture of GATOR

When parsing sentences provided by the designer, the Parser must be able to query the designer about any unknown words or terminology, both of which might be the result of misspellings. If a sentence is highly ambiguous, the Predicate Generator must explain the ambiguity to the designer so he/she can identify the appropriate interpretation. After parsing and interpretation, the sentences are converted into high-level predicates that are passed to the Command Interpreter. For example, if the designer entered the sentence:

"TWC allows a customer to add a third party to an existing conversation."

the following predicates is passed to the Interpreter:

```
(utterance   :type sentence :subtype SVO   :mood declarative
     :preds  (:subj (jargon-obj :name %t%w%c  :superclass process
                                :type telephony  :subtype feature
                                :mood direct    :syn-attr (GN N1))
             :verb %v%permit
             :obj (a-frame :form SVOLE
                     (%v%apply :verb (:name add%2 :tense future
                                       :voice active
                                       :pred (%v%append $subj $obj) )
                               :subj (:name customer :det indef
                                       :mode direct :type telephony
                                       :syn-attr (GX N1)
                                       :superclass (person organization)
                                       :subtype (%j%private %n%business))
                               :obj (:name party :det indef
                                       :mode direct :type telephony
                                       :syn-attr (GX N1)
                                       :superclass (person equipment)
                                       :subtype customer
                                       :attrib (ordinal 3))
                               :le  (:link %r%onto
                                       :env (:name conversation
                                              :det indef :type telephony
                                              :mode (existential direct)
                                              :syn-attr (GN N1)
                                              :superclass process
                                              :subtype %v%talk
                                              :attrib (%v%exist)) ) )) ))
```

After executing the knowledge base commands produced from the predicates, GATOR produces appropriate responses to the designer. These responses range from simple graphical or textual displays of information and queries asking for clarifications and additional information to paragraphs of text describing a feature. At all times the Response Generator must clearly identify to the designer what actions were taken by the system so he/she recognizes that the system properly interpreted the input.

2. *The Command Interpreter* (consisting of the Interpreter) interprets the predicates formed from the designer's input by the User Interface and issues appropriate update and queries to the Knowledge-Data Base. The Interpreter accepts this output from the Predicate Generator, determines what action is being specified, performs this action and any implied actions, and determines the appropriate high-level response returned to the user.

CONCEPTS

Hierarchy of Basic Knowledge Concepts

- Pre-constructed
- Built by System using Classification

Provides Structure for Instance Knowledge

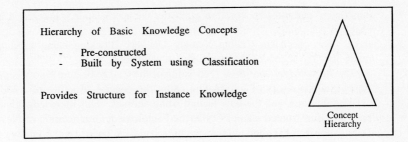

Concept Hierarchy

INSTANCES

Hierarchies of Knowledge Describing:

- Basic Call
- Various Features and Feature Classes

- Pre-constructed
- User-Defined

Call Description Hierarchy   Feature Hierarchy

MODELS

Constructed Model of a Call with Particular Feature(s) Used to:

- Reason About Call Structure
- Simulate Call Actions
- Develop Reqt. Document

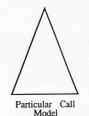

Particular Call Model

Figure 2. The Knowledge Levels within POND

Sometimes the output from the Predicate Generator specifies actions to be taken by the telephone switching circuits and software (e.g., "The Directory Number is always transmitted to the terminating office as a part of the Initial Address Message."). In other cases it provides structural/organizational knowledge (e.g., "Calling Number Delivery Blocking is a custom local area signaling services feature.") or describes actions for displaying information (e.g., "Display a call with Calling Number Delivery Blocking and Three-way Calling."), locating information within the representation (e.g., "What are the part of a call?"), creating new knowledge that must be stored (e.g., "After the access code is entered, it is checked for validity."), and inserting/deleting/moving existing knowledge (e.g., "The check for Calling Number Delivery Blocking being valid is made after the access code is verified as a valid code.").

3. *The Knowledge/Data Base* (consisting of the Call & Feature Knowledge Base) is a repository of information about the general structure of a call, existing features, and the new feature being defined. This knowledge/data base was constructed using POND (the Pantological Organization of New Delineations), a knowledge representation language based on the family of KL-ONE languages [13], which is discussed in the next section.

4. POND

KL-ONE is an Artificial Intelligence based knowledge representation language developed at Bolt Beranek and Newman Inc. by Ron Brachman based on ideas formulated in his Ph.D. dissertation [14]. Most KL-ONE languages partition knowledge into a Terminological Box (or T-Box) and an Assertional Box (or A-Box). The T-Box provides the structure and organization of the domain specific knowledge represented in the A-Box. For example, the T-Box might represent the concept of "a black dial phone" which describes any dial phone that is black. The A-Box states constraints or facts that apply to a particular domain where we might assert "This is John's black dial phone". POND differs from these languages by dividing knowledge into three rather than two levels (see Figure 2): Concepts, Instances, and Models.

4.1. Concepts

Concept-level knowledge is high-level, structural knowledge used to organize and configure all other knowledge within the knowledge base. This knowledge corresponds to the T-Box in a KL-ONE language and includes definitions of:

1. The concepts required to represent all the components of a telephone call, the various features, and feature classes. Figure 3 illustrates the definitions of two concepts, `top-level-task` and `task`. Each of these concepts contains several slots (e.g., `:features`) that define the type and number of values that they can contain, references (i.e., `:ako`, which denotes "a-kind-of") to other concepts on which they are based, and applicable constraints (i.e., `:annotation`).

2. Special slots, or attributes, of the concepts, which are slots that have a set of restricted values or that define relationships between concepts. For example, Figure 4 depicts the `category` slot, which has a set of restricted values, and the `children` and `parent` slots, which identify their interrelationship.

3. Temporal relations required within the concepts, which are used within Instances to specify temporal ordering. Figure 5 details `s`, the `starts` temporal relation. This definition states that `s` is a-kind-of (`:ako`) `time-relation`, is the inverse of the relation `si` (`starts-inverse`), and has certain restrictions (`:constraint`) on the starting (`start-time`) and ending (`end-time`) times of the tasks that are the `range` and `domain` of this relation.

```
(def-conceptq top-level-task
    (:feature children (:type task))
    (:feature start-time (:type integer)
                         (:number 1))
    (feature end-time (:type integer)
                       (:number 1))
    (:annotation :concept
       (:constraint (< start-time end-time)) ))

(def-conceptq task
    (:ako top-level-task)
    (:feature parent (:type top-level-task)
                     (:min 1))
    (:feature time (:type time-relation))
    (:feature cat (:type category)
                  (:number 1)
                  (:value comb)) )
```

Figure 3. The Definition of Two Concepts

```
(def-intervalq category
    (:value user sys comb timing usertimed))

(def-relationq children
    (:inverse parent))

(def-relationq parent
    (:inverse children))
```

Figure 4. Sample Definitions of Slots

```
(def-relationq s
    (:ako time-relation)
    (:inverse si)
    (:constraint
       (= (domain start-time) (range start-time))
       (< (domain end-time) (range endtime)) ) )
```

Figure 5. The Definition of the Temporal Relationship *s*

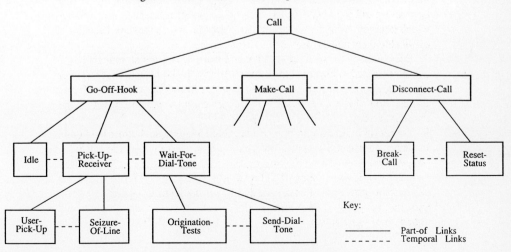

Figure 6. Partial Decomposition of *Call*

4.2. Instances

The instances consist of specific descriptions of the components of a telephone call, existing features, feature classes, and the new feature being defined. Instance knowledge corresponds to the A-Box knowledge in a KL-ONE language. Concept Level definitions form the structure for the knowledge defined on the Instance Level and provide some of the particular values possessed by the instances.

For example, Figure 6 contains a graphical representation of the instances that compose a *Call*. A *Call* initially decomposes into the instances of *Go-Off-Hook*, *Make-Call*, and *Disconnect-Call*. Each of these instances is further decomposed as shown. Note that explicit temporal relationships associated with each instance define the instance's location to the other instances in this decomposition.

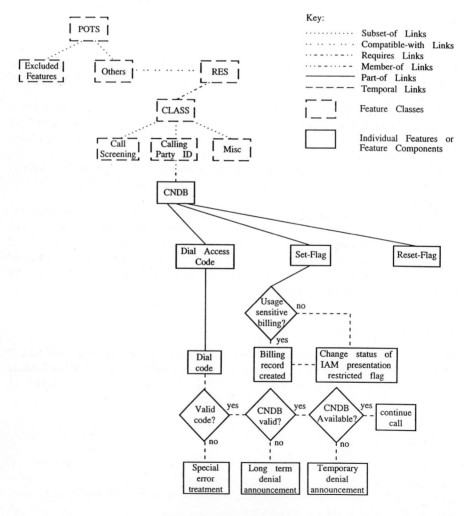

Figure 7. The Feature Hierarchy

The other portion of the Instance Level knowledge is a feature class hierarchy, which describes the relationship between the feature classes and individual features according to the type of telephone service required. As shown in Figure 7, *POTS* (plain, ordinary telephone service) is divided into features that are compatible with *RES* (residential services) and those that are not. *RES* provides an additional set of features beyond those provided by *POTS* and is required for *CLASS* (custom local area signaling services). *CLASS*, in turn, is divided into features involving *Call Screening*, *Calling Party ID*, and *Miscellaneous*.

Each individual features, such as Speed Calling, Last Number Redial, and Calling Number Delivery Blocking, is detailed within the feature hierarchy in a manner similar to the decomposition of *Call*. For example, Figure 7 includes a partial decomposition of Calling Number Delivery Blocking, *CNDB*. Each decomposition details the human actions that enable the feature, any resulting system actions, and the temporal relationships between the various components of the feature and the components of a *Call*.

4.3. Models

The Model Level allows the designer to construct a model of a call with various features to determine if GATOR has built an appropriate representation of the new feature on the Instance Level.

At any point during the process of describing a new feature, the designer can request a portrayal of a telephone call with this feature or some set of features. The system builds a model by examining its Instance Level knowledge and retrieving all the relevant instances (i.e., those representing a call and those representing the feature(s) of interest). The retrieved instances are then combined to create a model using any explicit or implicit interactions and dependencies that exist between the instances representing a call and the instances representing the feature(s). See Figure 8.

Once the model is created, it is presented to the designer for examination. Errors in the model result from incomplete or incorrect specifications on the Instance Level, so the designer must examine the model for correctness and must modify the Instance Level knowledge to correct any detected errors. Once corrections are made to the knowledge base, the designer can build another model and examine its structure to verify that appropriate corrections have been made.

5. Feature Interactions

The advent of ESS in the 1960s lead to the prolific increase in the number and diversity of telephony features. For GATOR and other related systems to be effective tools in assisting designers and developers, they must provide the ability to represent interactions that exist among features and must aid in identifying potential interactions.

5.1. Representing interactions

The utility of GATOR derives from the ability of the Instance Level knowledge to represent the three types of interactions that can occur between features. As discussed in Section 2, two features can be (1) incompatible, (2) compatible with special restrictions, or (3) compatible with no specified restrictions.

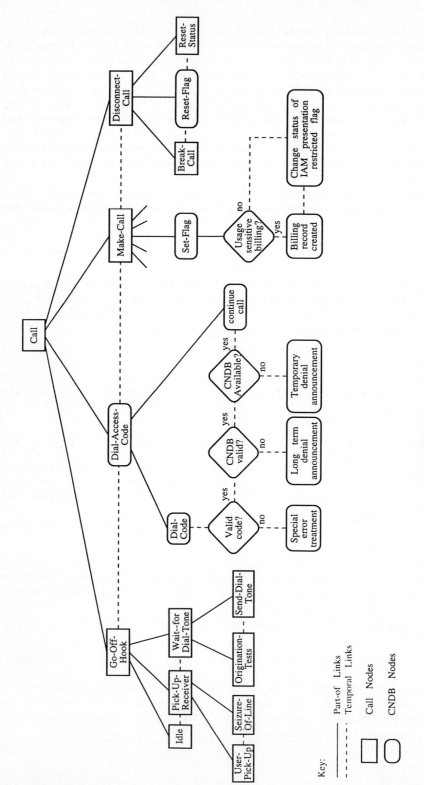

Figure 8. Model Representation of a *Call* with *CNDB*

Incompatibilities between two features are represented through values placed in a special slot contained in each feature description. When interactions exist, the feature hierarchy is not a tree but rather a graph. For example, consider Selective Call Forwarding (SCF) and Selective Call Rejection (SCR). Both features specify that the calling DN should be checked to determine if it is in a table of DNs and, if it is, appropriate actions are taken (for SCR the call is forwarded to the specified number and for SCR the call is rejected). Since both actions are specified to occur at the same time within a call there is a temporal conflict. In this particular case, SCR is designated to take precedence over SCF. A check is made to see if the call will be allowed to terminate at the called station. If it is accepted, the DN is then matched against the SCF screening list. See Figure 9.

When the designer requests that a model be created with some combination of features, the knowledge base is searched for explicit incompatibilities and interactions. For example, if the designer asks for a model of Three-way Calling with Speed Calling, GATOR will discover no explicit interactions so an appropriate model will be constructed based on each feature's characteristics (e.g., when the dialing of the third party is performed, it is done using Speed Calling).

Should such constraints exist (as is the case with SCF and SCR), the specification of these constraints is used when creating the model. Once the model is built and presented to the designer, it is then the designer's responsibility to determine if the model structure is correct, making any corrections or changes that are required.

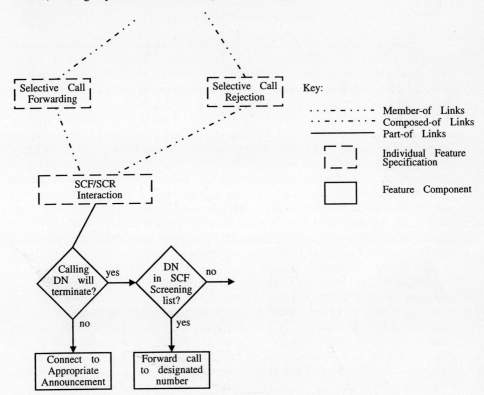

Figure 9. The Explicit Representation of the Interaction between Selective Call Forwarding and Selective Call Rejection

5.2. Discovering interactions

One of the most difficult tasks for a designer is determining the set of features with which a new feature interacts. Three forms of knowledge can aid this process. First, this task is aided by the organization of features into feature classes as shown in Figure 10, which details a portion of this feature class hierarchy. While this hierarchy originated from a need to detail what software is required to make particular features available (e.g., Calling Party ID features cannot function without the CLASS base software), it can also aid in determining potential interactions.

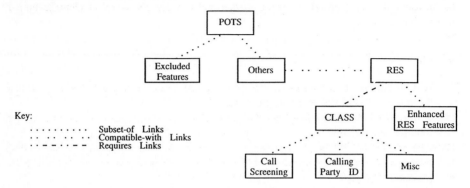

Figure 10. Feature Class Relationships

For example, suppose that the designer specified an interaction between some new feature and Calling Number Deliver Blocking (a member of the Calling Party ID feature class) when creating the description of the new feature. GATOR should assist the designer in determining what potential interactions exist between this new feature and the other members of the Calling Party ID class based on their shared

The second form of knowledge that can assist in determining feature interaction is inferential knowledge obtained from statements by the feature designer. Suppose that the designer has previously developed a description of the feature Calling Number Delivery. In this description, the designer identified that Calling Number Delivery is compatible with interactions to Call Waiting and Call Forward and is incompatible with Denied Termination (this feature allows a telephone company to deny all call terminations to a customer's line at the customer request). When developing a description of Calling Name Delivery the designer might state "Calling Name Delivery is just like Calling Number Delivery except the calling name is delivered to the called party rather than the calling number." An effective tool should use this knowledge to modify the description of Calling Number Delivery to create a description of Calling Name Delivery or, minimally, to suggest potential interactions for Calling Name Delivery based on the interaction that exist for Calling Number Delivery.

The final form of knowledge that can assist in determining feature interactions is case-based knowledge. As knowledge of various features is developed within a tool, similarity measures among the features can be computed to categorize their resemblance to each other. When the initial description of a new feature is found analogous to an existing feature, the existing feature's interactions can be used as a guide for potential interactions of this new feature.

Note that each of these approaches provides an initial set of potential interactions that should be examined by the designer. These potential interactions do not in any way preclude the existence of others nor should they be viewed as anything more than potential interactions that should be confirmed with the designer.

6. Development Status

The current GATOR prototype, while limited in its natural language capability, can create new features and their sub-components, can answer questions about a feature's structure and organization, and can create models detailing how features combine with the description of a call. Our research efforts are currently concentrating on (1) improving the language capability of the system through an expansion of its grammar and lexicon, (2) developing output generation facilities for the production of paragraph and technical report length descriptions, (3) expanding the knowledge base of telephony features, and (4) developing a case-based reasoning system and other tools to aid in identification of feature interactions.

References

[1] G. E. Schindler, editor, A History of Engineering and Science in the Bell System, Switching Technology (1925 - 1975), Bell Telephone Labs, 1982.

[2] Northern Telecom, *Northern Telecom Software*, 5204.11/11-89, Issue 1, Northern Telecom, Research Triangle Park, NC, 1989.

[3] A. Ranade, and J. C. Petrides, "A Pragmatic View of Feature Interaction Management", *International Workshop on Feature Interactions in Telecommunications Software Systems*, St. Petersburg, FL, December 1992, pp. 13 - 21.

[4] D. G. Belanger, S. G. Chappell, and M. Wish, "Evolution of Software Development Environments" *AT&T Technical Journal*, March/April 1990, pp. 2 - 6.

[5] D. G. Belanger, R. J. Brachman, Y. Chen, P. T. Devanbu, and P. G. Selfridge, "Toward a Software Information System" *AT&T Technical Journal*, March/April 1990, pp. 22 - 41.

[6] P. Devanbu, R. J. Brachman, P. G. Selfridge, and B. W. Ballard, "LaSSIE: A Knowledge-Based Software Information System", *CACM*, Vol. 34, No. 5, 1991, pp. 34 - 49.

[7] G. D. Bergland, G. H. Krader, D. P. Smith, P. M. Zislis, "Improving the Front End of the Software-Development Process for Large-Scale Systems" *AT&T Technical Journal*, March/April 1990, pp. 7 - 21.

[8] V. Kelly, and U. Nonneman, "Inferring Formal Software Specifications from Episodic Descriptions", *Proceedings of AAAI-87*, AAAI Press, Menlo Park, CA, 1987, pp. 127 - 132.

[9] ISO, IS 8807, *Information Processing Systems - Open Systems Interconnection - LOTOS - A Formal Description Technique based on the Temporal Ordering of Observational Behaviour*, The International Organization for Standardization, May 1989.

[10] W. L. Johnson, "Specification as Formalizing and Transforming Domain Knowledge", *Proceedings of the Workshop on Automating Software Design*, AAAI-88, St. Paul, MN, August 1988, pp. 48 - 55

[11] Reasoning Systems, "Use of REFINE™ for Communication System Design, Analysis, and Synthesis", Application Note 1.2, Reasoning Systems Inc., Palo Alto, CA, August 1986.

[12] D. D. Dankel, W. Walker, M. Schmalz, and K. Nielsen, "A Model for Capturing Requirements", *Toulouse '92, Software Engineering and its Applications*, EC2, Nanterre Cedex, France, December 1992, pp. 589 - 598.

[13] W. Woods and J. Schmolze, "The KL-ONE Family", TR-20-90, Center for Research in Computing Technology, Harvard University, Cambridge, MA, 1990.

[14] R. J. Brachman, "A Structural Paradigm for Representing Knowledge", Ph.D. Dissertation, Harvard University, May 1977.

Author Index

Atlee, J.M.	36	Lin, F.J.	86
Blom, J.	197	Linden, R. van der	24
Braithwaite, K.H.	36	Logrippo, L.	136
Cameron, E.J.	1	Magill, E.H.	236
Cheng, K.E.	152	Muller, J.	73
Combes, P.	120	Muzzi, L.	258
Cross, M.	249	Nielsen, K.	258
Dankel, D.D.	258	Nilson, M.E.	1
Faci, M.	136	O'Brien, F.	249
Gammelgaard, A.	178	Ohta, T.	60
Griffeth, N.D.	1,217	Pickin, S.	120
Harada, Y.	60	Rhodes, D.	258
Jonsson, B.	197	Schmalz, M.	258
Kempe, L.	197	Schnure, W.K.	1
Kimbler, K.	73,167	Søbirk, D.	167
Kristensen, J.E.	178	Tsang, S.	236
Kuisch, E.	73	Velthuijsen, H.	1,217
Lin, Y.-J.	1,86	Walker, W.	258